T0353567

Next Generation Safety Leadership

Next Generation Safety Leadership

From Compliance to Care

Clive Lloyd

CRC Press

Taylor & Francis Group

Boca Raton London New York

CRC Press is an imprint of the
Taylor & Francis Group, an **informa** business

First edition published 2020
by CRC Press
6000 Broken Sound Parkway NW, Suite 300, Boca Raton, FL 33487-2742

and by CRC Press
2 Park Square, Milton Park, Abingdon, Oxon, OX14 4RN

© 2021 Taylor & Francis Group, LLC
CRC Press is an imprint of Taylor & Francis Group, LLC

Library of Congress Cataloging-in-Publication Data

Names: Lloyd, Clive (Clive F.), author.
Title: Next generation safety leadership : from compliance to care / Clive
Lloyd.
Description: First edition. I Boca Raton, FL : CRC Press, 2020. I Includes
bibliographical references.
Identifiers: LCCN 2020020830 (print) I LCCN 2020020831 (ebook) I ISBN
9780367509538 (hardback) I ISBN 9781003051978 (ebook)
Subjects: LCSH: Industrial safety--Psychological aspects. I Employee
motivation. I Corporate culture. I Leadership. I Trust.
Classification: LCC HD7261 .L535 2020 (print) I LCC HD7261 (ebook) I DDC
658.4/08--dc23
LC record available at https://lccn.loc.gov/2020020830
LC ebook record available at https://lccn.loc.gov/2020020831

ISBN: 978-0367-50953-8 (hbk)
ISBN: 978-1-003-05197-8 (ebk)

Typeset in Times
by Deanta Global Publishing Services, Chennai, India

CONTENTS

PART 2

FOREWORD

Safety as an industry is in crisis.

For reasons that escape me, Australia seems to be clinging to out-dated, culture-eroding, fear-driven 'traditional' safety models and theories that we just can't seem to shake. We have trivialized what it means to 'be safe,' adopting punitive approaches to safety built upon weak foundations of words on paper, tick-and-flick exercises, PPE, and 'catching someone doing something wrong.' And the scary thing is that the majority of managers, people leaders, and even safety professionals I come across genuinely believe this approach is the right one.

Sometimes I catch myself wondering if I want to remain part of something that just doesn't seem capable of changing for the better. But then I shake it off by reminding myself of why I started in the field of safety in the first place, taking inspiration from some of the great people effecting positive change like Clive.

You might be thinking this all sounds a little grim – quite a negative tone to open with. And you would be right – it is. But I am an honest person, and I have learned that the best way to approach a challenge is to stare into the brutal facts in order to establish exactly what it is I am trying to fix. Yes, I believe that safety as an industry is in crisis. But I also believe there are still opportunities to drag our out-dated notions of safety management into the modern era. And here's how I am going about it, employer-by-employer, assignment-by-assignment: After all, you can only eat an elephant one bite at a time:

- We need to move away from KPIs and toward a state where we embrace the human condition and recognize its positive contribution to safety. We need to get better at harnessing human potential to see people as the solution, not the problem. After all, human error is not 'the cause'; it is a symptom of something much bigger. In Clive's words, behaviors are not the problem; they are expressions of the problem.
- We should be looking to high reliability organizations (HROs) for inspiration on how to create states of collective mindfulness and how to build resilience and therefore capability.
- We should be sharing with our leaders and managers the great conversations that are being had in progressive safety forums around concepts such as Safety II and Safety Differently – or even better, dragging our leaders along to these forums to engage and be engaged directly.
- We need to recognize trust as a key predictor of mature safety cultures, build psychologically safe workplaces where positive safety interventions are embedded as blueprints in our subconscious library, and view our employees as customers of our safety programs.
- We need to engage with the workforce and listen to learn, not listen to respond. Many of the problems we are trying to address are due to our inability to listen.
- And perhaps most importantly, we need to be outwardly curious, empower ourselves and others to share stories about safety, and learn from each other, because I am certain that someone out there has faced the same safety challenges I am facing and found effective ways to overcome them.

Safety as an industry is in crisis. Beginning the shift from compliance to care through next generation safety leadership as outlined here by Clive may just provide some answers to questions that not enough of us are asking.

Tim D'Ath
Head of Safety & Environment
Melbourne and Launceston Airports

PREFACE

Now well into my third decade as a psychologist – most of that time specializing in safety leadership – I'm heartened by the earnest efforts of researchers who continue to identify key predictors of workplace safety culture and performance. I am also perturbed by how few industry leaders put such evidence-based findings into practice.

Generally, each successive generation builds upon the lessons of the past. This, however, has not necessarily been the case in the safety field. The top-down, hierarchical model of seeking to force compliance is as prevalent now as it was 20 years ago, which tends to be reflected in the plateau many companies are experiencing in their injury rates.

WHY IS THAT?

In a recent conversation, Dr. David Provan (research fellow with Griffith University) suggested to me that "Safety Science as a discipline is about 30 years behind its parent disciplines, such as psychology."

I tend to agree – but again, why?

Safety leaders are not (usually) academics who spend their precious time between mountains of paperwork and those infernal 'safety walks' reading peer-reviewed research (although I know a few that do). They have many other competing priorities and KPIs to deliver upon – few (if any) of which would include asking the question, "is what we're doing actually evidence-based?" Busy safety leaders are not necessarily interested in the statistical analyses behind a study – they don't care if $p < .05$ – they just want to know what the relevant research means for them and how they can use it to keep their people safe.

Moreover, much of the current research into effective safety leadership (and, indeed, leadership, in general) points to the importance of 'soft skills' (a misleading descriptor), which the *Oxford English Dictionary* defines as

> Personal attributes that enable someone to interact effectively and harmoniously with other people.

Much of my work over the last two decades has been in the mining, oil, and gas and construction sectors, comprising organizations I tend to label as 'macho' (disproportionally male, 'old school,' etc.). In these industries, quite often, soft skills tend to be viewed with the same disdain as yoga, tofu, and leafy green vegetables! Leaders in these organizations (it would seem) would much prefer to just stick with engineering controls and safety management systems than entertain all this touchy-feely psycho-babble!

So it's frequently a case of "we don't know what we don't know," and even if we do know about this stuff it sounds hard (and paradoxically 'soft' at the same time!). Little wonder then that safety leaders revert to doing what is comfortable. Sadly, this is often based on seeking to force compliance via methods mired in out-dated modalities such as behaviorism. If technology had developed at the same rate as safety we'd still be using Commodore 64s and dot matrix printers!

This is both frustrating and understandable, and while it may be tempting to chide safety leaders for their apparent reluctance to move toward evidence-based practices, perhaps psychologists and researchers in associated areas need to get better at providing leaders with clear and tangible strategies, rather than the mere exposition of abstract theories about triangles, bow ties, and slices of Swiss cheese.

The recent ascent of the Safety Differently movement is a welcome response to the lack of fundamental change in how safety has been framed by generations of organizational leaders and the safety profession in general. The fundamental premise of Safety Differently (i.e., people are the solution, not the problem) promises a more humanistic trajectory after decades of safety resembling boot camp, yet at present, in the minds of many in the safety field (if not academia), Safety Differently remains more as an appealing philosophy than as an implementable approach with compelling research evidence attesting to its efficacy.

There is, however, a strong and consistent predictor of safety performance that is readily available to be harnessed by all organizations and leaders, one that any safety initiative depends upon for success, and yet one that has largely been overlooked by the safety profession. This is a book about trust.

Without trust (and its cousin, Psychological Safety):

- People don't speak up.
- Teams don't fully engage.
- Metrics don't reflect reality.
- New initiatives don't gain traction.
- People don't admit mistakes.
- People don't share ideas.

What is needed in the safety field is a fundamental shift – generational change – where leaders seek to create trust first and have the permission, knowledge, skills, and abilities to do so.

The book draws widely from the relevant literature, yet it is not solely based on academic research. After spending 20 years in the field, working with executives, senior leaders, and frontline staff around the globe and across many industries, I have learned a lot about what great (and not so great) safety leadership looks and sounds like in practice, and I'll be sharing examples of both.

This is a book for leaders, from the boardroom to the frontline. If you are an executive and believe that safety is best left to the 'safety team,' think again! If CEOs feel the board will not support a fundamental shift, they will be unlikely to pursue one, and that in turn will inhibit change throughout the organization. It starts at 'the top,' or it doesn't start at all!

In the coming chapters we will explore what works (as well as what doesn't) in terms of creating a climate where trust and psychological safety can thrive.

Part 1 of the book reviews the research into safety culture development with a particular focus on the crucial role of psychological safety. I then discuss how some common safety leadership philosophies, tools, and language can inadvertently reduce trust and impede safety culture development.

Part 2 offers the reader practical skills and strategies that can be used to build trust and psychological safety. Their consistent application will assist your organization's safety culture to mature from one of rule-based compliance to an inclusive climate of care.

HOW TO USE THIS BOOK

As you read through the book, please take note of the reflection questions at the end of each chapter. I strongly recommend you write your thoughtful responses to these questions in a separate journal, as your answers will form the basis for some powerful activities you can subsequently facilitate with your work team and leadership group – these activities will be discussed in Part 2.

Once you've finished the book, I recommend you then return to specific chapters – particularly those in Part 2 that emphasize the development of particular skills and strategies. Focus on one chapter at a time and seek to put a key strategy or skill into practice. Once you start to gain some degree of proficiency in a given area, proceed to the next chapter and repeat the process. Gaining mastery in these areas will ensure you, as a leader, are well equipped to create the trust within your team members that will enable them to thrive in a collaborative climate of high psychological safety.

ACKNOWLEDGMENTS

To me, luck is what happens when preparedness meets opportunity. I had been meaning to write this book for a while, yet finding the time required to even begin (let alone complete) such an undertaking seemed like a fantasy. Then I received a message from a guy I'd never met in person but occasionally interacted with on the LinkedIn platform. He asked me if he could use one of my (frequent) quotes about behavior based safety in his book. That led to further discussions, and before I knew it his publishers were asking if I would review his manuscript – which I did.

After submitting my review, his publishers contacted me again and informed me that when I was ready to write my own book they would be happy to publish it ... nice! That was the opportunity and catalyst I apparently required to start writing this volume. So I would like to thank Jason Maldonado, author of the excellent *A Practical Guide to the Safety Profession* for inadvertently inspiring me to stop making excuses, get off my butt, and get this book done! Thanks also to CRC Press and the Taylor & Francis Group for the offer to publish it, and especially to Michele Smith and Shannon Welch, for their valuable assistance throughout the project.

I have been fortunate over the years to have worked with and learned from some outstanding safety leaders, and many of them directly or indirectly contributed to this book. Without exception they fulfill the criteria for what I will go on to describe in the book as 'Next Generation Safety Leaders,' and many were early adopters of my approach. As such they showed a degree of courage by being willing to follow a maverick psychologist who dared break free of the behaviorist's 'black box' to explore a whole new world of safety leadership possibilities grounded in the more fertile modalities of cognitive, social, humanistic, and positive psychologies. While far from an exhaustive list, some of these luminaries include: Richard Luxton; Nick Fitzpatrick; Sophie Grace; Anthony Donald; Mario Russo; Deonne Drew; David Spackman; Jo Christofides; Adam Mroz; Jacqui Higgins; Jim Miller; Bryan Fuller; Peter Shaw; Ronald de Geus; Han Poort; James Hall; and the outstanding Tim D'Ath (who was kind enough to write the foreword).

John Phillips and Simon Bown are also exemplars of the next generation safety leaders, and you'll read about both of them in Part 2 of the book.

My 'Brain's Trust' network of connections on LinkedIn has also been very generous with their time and expertise, supplying me with insightful responses to tricky questions I posed from time to time – thank-you one and all!

To all our Care Factor Program participants over the years, whether you are board members, senior or frontline managers, or members of work teams. I have learned from each and every group – thank-you for your courage, candor, and willingness to 'let me in.'

Many thanks to my partner, Tanya. As well as being my partner in life, she is my partner in business, and while my myopic attention was on writing, she was left carrying the load in terms of keeping our company, GYST Consulting, going. She was also a consistent voice of rationality when I wished to throw my laptop in the bin!

To our team at GYST Consulting, thank-you for your support, encouragement, and coaching, and for consistently demonstrating the very leadership attributes I describe in the coming chapters – I am fortunate indeed to be working with such a professional, knowledgeable, and brilliant group of people, and similar expressions of gratitude have been consistent themes in feedback from our clients.

Thanks also to my fellow psychologists Dr. Luke Daniel and Jamie Toth, for providing feedback on the initial proposal, and to Angela Maggacis from Mopki, for providing illustrations that bring to life some of the book's key concepts.

It would be remiss of me if I didn't also acknowledge my proofreader-in-chief and father-in-law, Eric Pavey, who, at age 83, cast his eagle eyes over the manuscript and found me wanting (and

seemed to take an inordinate amount of pleasure in doing so) –
thanks Eric!

Finally, thanks to you, dear reader! I hope you find this journey
from compliance to care interesting, thought provoking, useful, and
enjoyable.

ABOUT THE AUTHOR

Clive F. Lloyd is an Australian-based Psychologist specializing in Safety Leadership and Culture Development. He was recently named among the top 50 global thought leaders and influencers on culture by Thinkers360. He is co-owner of, and Principal Consultant with, GYST Consulting Pty Ltd and developer of the acclaimed Care Factor Program. Clive has spent the last 20 years assisting organizations to improve their safety performance by developing trust and psychological safety and doing Safety Differently. He has worked with global mining, oil, and gas, construction, and utilities companies in Australia, New Zealand, Fiji, USA, Canada, South Africa, Norway, Singapore, China, UAE, PNG, KSA, United Kingdom, and Costa Rica. He lives on Australia's Gold Coast with his wife, Tanya, and two rescued working dogs.

Clive can be contacted via:

Website: www.gystconsulting.com.au

Email: info@gystconsulting.com.au

gystconsulting°

The home of psychological safety

GROW
YOUR
SAFETY
THINKING

PART **I**

TRUST
The Currency of Leadership

Trust arrives on foot but leaves on horseback

– Dutch Proverb

When we're facilitating workshops for teams, we request (whenever possible) that a senior leader opens up the day. This shows that the leadership team is committed to and invested in the change process. Similarly, we ask that, whenever possible, a senior leader comes in at the end of the session to close out and perhaps take a few questions from the group based on topics that may have arisen during the workshop.

This approach generally works very well, however, not always!

In a recent session at a mine site, a Safety Superintendent, 'Greg,' opened the session and did a great job. Greg had already been through the program and was looking forward to his team gaining similar insights. He thanked everyone for their attendance and suggested the training was part of the company's journey toward doing *Safety Differently.*

The session went well, and the group was animated, curious and candid throughout the day. They came up with many action items for themselves and a few probing questions to ask Greg when he returned to close out the session. Except it wasn't Greg who came (he was otherwise occupied) – it was the HSE Manager – 'Nigel.' I had not met Nigel prior to this point as he had been on long service leave and, hence, had not participated in the program. He began, not by enquiring about the group's experience of the day, but by saying the following (and this is pretty much verbatim).

So, I hope you got a lot out of the group – this guy is not cheap! I'm looking forward to seeing our incident numbers coming down. I hope you now see the importance of complying with the safety 'golden rules' and why I have a zero tolerance for rule violations. So, are there any questions for me?

Despite the fact that there was a list of questions on flip chart paper very close to where Nigel was standing, nobody in the group asked him anything. Some participants were looking down at the table; others were looking at me – *nobody* was looking at Nigel! His final offering was, "Gee Clive, tough crowd! Looks like you've bored them stupid!"

Suppressing and editing the unsavory words that first came to my mind, I thanked Nigel for stepping in for Greg and let him know the group and I still had a few things to discuss before wrapping up. I added that I'd come over and chat with him later (which I did, and the 'chat' quickly morphed into a much-needed coaching session!).

Inadvertently, what Nigel had done in those few, excruciating minutes was essentially to undermine any fleeting hope the attendees may have once possessed that anything was about to change in terms of the team culture. He had raised the fear level about speaking up and consequently reduced the level of trust within the team. The unintended costs of Nigel's 'pep talk' illustrate the fragility of trust. It can take a long time to build, yet be destroyed in seconds.

My experience, as well as the research, tells me that where fear is high, trust and psychological safety will be low (often indicated by an employee's reluctance to speak up). Conversely, in a high trust culture, such fears are no longer present and people feel free to share their concerns and ideas. As Edmondson (2019) warns, "no twenty-first century organization can afford to have a culture of fear" (p. xix).

Similar observations were shared by Professor Patrick Hudson, who identified *increasing levels of trust* as an indicator of maturity in his *Five Levels of Safety Culture Model* (Hudson, 1999, 2001). The original model is now almost 20 years old, and while there have been a number of variations to this framework over the years (including by the Keil Centre which used slightly different terminology to describe its *Safety Culture Maturity Model*), it still holds up quite well.

Nevertheless, while drawing upon more current research, and with a particular focus on the social–psychological aspects of safety culture development, I found it useful to propose a revised version of Hudson's original model, presented in Figure 1.1.

In Hudson's original model, the levels were drawn as separate, discrete entities, suggesting that a company is at one level or it's not. The

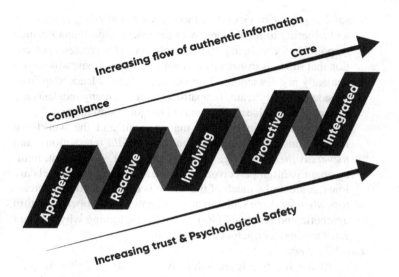

Figure 1.1 Mapping the journey from compliance to care.

revised model shows level progression more as a journey, including the inevitable ups and downs encountered along the way.

Hudson's model also listed *an increasingly informed workforce* as an indication of progress; however, that still infers top-down, one-way communication. I have revised that to *an increasing flow of authentic information*, which points to the fact that leaders of more mature cultures don't merely keep their employees well informed – *rather they are also well informed by their employees* – indicating *genuine two-way communication*. In order for a team to feel comfortable speaking up, *psychological safety* must be present; hence it has also been added to the model.

Finally, the essence of the revised model depicts a journey from *compliance* to *care*, and illustrating what that means is the central theme of this book.

A brief description of each level's basic attributes is presented below:

Level 1 – Apathetic

In apathetic cultures, management adopts a 'blame the worker' approach in that incidents are generally seen as a result of a worker's stupidity, inattention or willful violation. 'Being safe' is primarily viewed as mechanically following procedures and adhering to regulations, with the safety department deemed responsible for 'policing' such compliance. This creates a perception that safety is distinct from day-to-day operations, which conveniently negates any need for visible, felt safety leadership from outside the safety team. In apathetic cultures, many incidents are seen as unavoidable and just part of the job ("sh#t happens!").

Communication between management and the workforce largely consists of top-down *parent-to-child* interactions, and *us versus them* language is highly prevalent. As a result, management is often perceived to be uncaring, and trust levels are low (incidentally, much of the above is experienced by contractors when working with client organizations that operate within apathetic and reactive cultures, partly explaining why incident rates among contractors are so high).

Level 2 – Reactive

At the reactive level, safety is a priority ... *after* an incident! Senior managers may apply elements of behavior-based approaches (e.g., punishment) when incident rates increase and may operate under the errant assumption that the majority of incidents are *solely caused by the unsafe behavior of front-line staff.* Hence, among the workforce there is still a degree of fear around reporting incidents, and secrets are often kept from management, impeding the authentic flow of potentially vital information. Unsurprisingly then, reactive organizations tend to have more than their share of serious incidents (Hudson, 2001).

Level 3 – Involving

Companies operating at the involving level recognize that the active participation of the workforce in safety discussions is important; hence teams are invited in to contribute. Consequently, as trust and psychological safety increase, employees become more willing to work with management to improve health and safety. Moreover, leaders are now prepared to concede that a wide range of factors cause incidents including *management decisions.* Safety performance is actively monitored, and the data is used purposefully. The organization has developed systems to assist with hazard management; however, the systems are often rigidly applied (Hudson, 2001).

Level 4 – Proactive

At the proactive level, the majority of employees in the organization believe that health and safety is important from both an ethical and economic point of view. Leaders and staff recognize that a wide range of factors cause incidents and the root causes are *likely to come back to management decisions* (Hudson, 2001). There is a growing recognition around the importance of all employees feeling valued and being treated respectfully, which helps build trust and psychological safety. The *us versus them* language associated with less mature levels is replaced by *we*, and communication between management and the workforce increasingly consists of two-way *adult-to-adult* interactions. The organization puts significant effort into proactive measures to prevent incidents through visible, felt safety leadership and by demonstrating genuine care for its people. Safety systems are designed to support staff, not the other way around.

Level 5 – Integrated

At the integrated level, leaders have fully invited their teams in, as they are seen as the subject matter experts. Leaders have created the climate necessary (high trust and psychological safety) for the workforce to accept responsibility for managing their own risks.

Safety is not viewed as 'separate' from the work done – safety is just how the organization does business, and the focus is on reliability, learning and doing work well. While such organizations may have had a sustained period (often years) without a recordable or high potential incident, there is no feeling of complacency. They live with the knowledge that their next incident is just around the corner, yet they are highly resilient when dealing with challenges (Hudson, 2001). The organization uses a range of indicators to monitor performance, but it is not performance driven, as it has *trust in its people and processes*. As a learning organization, it is constantly striving to improve and find better ways to design and implement hazard control mechanisms with the full involvement of the workforce (Hudson, 2001).

I have worked with many global multi-site corporations, and it is interesting to note that they tend to use the same procedures, policies, equipment, PPE, etc., across their various sites, yet one site can be operating at the reactive level and another at the proactive level – so why the difference?

Leadership!

Edmondson (2019) found that psychological safety differed substantially between groups. She noted that the leaders of some groups had been able to effectively create the conditions for psychological safety to thrive, while other leaders had not. In the more mature safety climates (proactive/integrated), it is not *golden rules*, policies, procedures, safety signs or slogans that increase in frequency or significance – it is trust, visible felt leadership, psychological safety and authentic information sharing that make the difference.

Unless these trust-related cultural aspects are attended to, an organization simply cannot progress to the more mature levels, regardless of the quality of safety management systems in place.

It is my belief that trying to change a safety culture by focusing *directly* on the culture is virtually impossible. Most organizational cultures are too big, too old, too intangible and too unwieldy to address directly. Little wonder then that many organizations feel dejected and dispirited when their tenth culture change campaign fails. Learned helplessness kicks in, and they simply give up, resulting in an inevitable return to the stick and carrot approach.

As Professor Andrew Hopkins (2005) stated,

> the evidence is that safe behavior programs do not work when the workforce mistrusts its management. Where such beliefs prevail, employers must first win the trust of their workforce by tackling some of the issues they see as affecting safety.

Professor Hopkins' thesis was supported by an Australian study focused on the coal mining industry. Gunningham and Sinclair (2012) found that "unless the mistrust of the workforce can be overcome then *even the most well-intentioned and sophisticated management initiatives will be treated with cynicism and undermined.*"

Think about that! Even the more forward-thinking, early adopters of *Safety Differently* approaches are likely to experience cynicism and pushback from the workforce unless they first overcome any existing mistrust among the workforce. 'Nigel' has some work to do!

First, create trust!

The same authors went on to conclude that

> Without trust workers treated almost all corporate management safety initiatives with suspicion and refused to buy into them. Safety observations were perfunctory, incident reporting was trivialized or ignored, and sophisticated electronic monitoring systems were side tracked. Second, mine management leadership on OHS (and middle management commitment) is every bit as important as corporate leadership and worker mistrust will not be overcome without it. Indeed,

the majority of workers does not identify with corporate management, and are largely unaware of them. This emphasises the importance of personal relationships and daily contact with mine site management in providing the opportunity for the demonstration of safety leadership. Accordingly, developing mechanisms that build a cooperative relationship between mine site management and the workforce and obtain worker 'buy-in' to management based initiatives will be crucial.

Similarly, the companies I work with that have been successful in moving toward proactive or integrated cultures have been those whose leaders *focus on relationships* rather than rules and enforced compliance – they *engage* with their people, building psychological safety and trust.

A plethora of studies have identified trust as a key predictor of safety performance and an essential component of proactive safety cultures (e.g., Carrillo, 2020; Edmondson, 2019; Burns et al., 2006; Eid et al., 2012; O'Dea & Flin, 2001).

Specifically, findings from these studies show that trust in management can increase employee engagement in safety behaviors and reduce the rates of accidents (Zacharatos et al., 2005). Conversely, other studies noted that mistrust is associated with diminished personal responsibility for safety (Jeffcott et al., 2006) and increased injury rates (Conchie & Donald, 2006, cited in Conchie et al., 2011).

Most companies want a culture in which people are willing to report hazards and near misses. Yet all too few create the psychological safety and trust required for such behaviors to become the unequivocal norm. Conchie et al. (2011) stated that

> trust is the foundation of a successful reporting program, and it must be actively protected. Even after many years of successful operation, a single case of a worker being disciplined as the result of a report could undermine trust and stop the flow of useful information.

Carrillo (2020) also noted that without trust, it is almost impossible to get reliable and timely information, especially if someone has made a mistake.

In short, trust is the primary currency for leaders. Without it, nothing else you do will make much difference. The next generation of safety leaders will be acutely aware of how their day-to-day interactions influence the trust and psychological safety of their people and will possess the knowledge, skills and desire to modify their leadership style accordingly.

PSYCHOLOGICAL SAFETY

In her now-classic work *The Fearless Organization*, Amy Edmondson (2019) defined psychological safety as

> a climate in which people are comfortable expressing and being themselves. More practically, when people have psychological safety at work, they feel comfortable sharing concerns and mistakes without fear of embarrassment or retribution. They are confident that they can speak up and won't be humiliated, ignored or blamed. They know they can ask questions when they are unsure about something. They tend to trust and respect their colleagues.
>
> (p. xvi)

While acknowledging that trust and psychological safety are strongly related, Edmondson (2019) distinguishes between these variables, suggesting that the latter is experienced at a group level, whereas trust refers to interactions between two individuals or parties. Despite this distinction, Edmondson (1999) also recognized the importance of trust in the development of psychological safety and noted that trust between team members is closely associated with members' psychological safety (Edmondson, 2003).

TRUST

> There is one thing that is common to every individual, relationship, team, family, organization, nation, economy and civilization throughout the world – one thing that which, if removed, will destroy the most powerful government, the most successful business, the most thriving economy, the most influential leadership, the greatest friendship, the strongest character, the deepest love.
>
> On the other hand, if developed and leveraged, that one thing has the potential to create unparalleled success and prosperity in every dimension of life. Yet, it is the least understood, most neglected, and most underestimated possibility of our time. That one thing is trust.
>
> – Stephen M.R. Covey (2006, p. xxvii)

The above citation from Covey's *The Speed of Trust* contains some rather big claims and on the surface could be read as hyperbolic, yet I believe few would disagree with them. Most of us can testify through lived experience to how a loss of trust can deeply impact our relation-ships, both inside and outside our organizations.

We all experience trust (or mistrust) in terms of our beliefs about the reliability, honesty and/or ability of someone. Trust is primal and fundamental. The decision to trust (or not) can be a life or death prop-osition. As such, the emotions associated with trust (and particularly mistrust) can also be primal and fundamental. In the workplace, the anxiety, resentment or deep-seated fear a person feels due to mistrust can make the difference between:

- Speaking up or remaining silent
- Doing a job the *right way* or the quick way
- Admitting mistakes or hiding them
- Engaging in safety initiatives or undermining them
- Sharing ideas or holding them in

It is not difficult to see why companies that have high levels of mistrust among the workforce hurt more people and tend to languish around the lower levels of safety climate maturity. Despite this, many organi-zations (and many leaders) seem reluctant to address this seemingly obvious challenge by making concerted efforts to build and sustain psychological safety. Given the likely payoffs listed above, one would imagine the 'why' for such a worthy goal would be highly apparent to most leaders. I believe any aversion is more about the 'how.'

Recent research has shown that by focusing on strategies that improve trust across all team members, psychological safety can be improved (Triplett & Loh, 2017), and these strategies, dear reader, will be a key theme in Part 2 of this book.

Let's begin this journey by getting clearer about how employees make decisions about the trustworthiness of a leader.

TRUST – A WORKING MODEL

One of the most frequently cited models of trust (particularly in the literature around leadership and organizational culture) was posited by Mayer et al. (1995). This integrated model suggests that trust is based on a worker's perceptions about three key factors:

1. Ability (perceived competence)
2. Integrity (perceived honesty and openness)
3. Benevolence (perceived degree of care shown) (Figure 1.2)

Figure 1.2 Three-factor model of trust.

Ability

The ability factor is reflected in how a leader's team perceives his/her levels of competence. While leaders need to demonstrate ability in required areas, no team expects its leader to be great at everything. However, if the leader has claimed a high degree of competence in a given area yet fails to live up to such claims, mistrust can be the result. If a leader's CV was inflated with hyperbole and exaggeration, she may be setting herself up for a fall. Similarly, if a leader believes he has the answers to all the team's questions (nobody does!), the team will soon work out he is full of ... it! (There's a lot to be said for under-promising and over-delivering!).

Next generation safety leaders are more authentic in their claims. When they don't have the answers, they are humble enough to say so and commit to finding them and/or working through issues *with* their teams.

Integrity

The integrity factor is reflected in how a leader's team perceives her/his honesty, uprightness, reliability and consistency. When leaders 'walk the talk,' follow through on commitments, keep their teams well informed and operate based on values rather than political expediency, they build trust within their teams. Conversely, if a leader plays favorites, fails to follow through on commitments, withholds information, demonstrates dishonesty, criticizes others in their absence or uses manipulative language, the subsequent loss of trust can be rapid and severe.

Next generation safety leaders are values based and use their organizational and personal values as decision-making tools that enable consistency, reliability and the building of psychological safety.

Benevolence

The benevolence factor is reflected in the degree to which a leader's team experiences him/her demonstrating care, empathy, compassion or kindness. Simply listening to an employee's concerns and acknowledging a job well done are tangible expressions of care. It can be demonstrated by *visible felt leadership* out in the field and by displaying a genuine interest in people.

Next generation safety leaders understand that showing care is not a sign of weakness that somehow threatens or negates their power, status or (in some cases) manhood. They exhibit empathy combined with outstanding communication skills (which will be a focus area in Part 2).

To paraphrase Meatloaf, "two out of three 'aint bad" – however, when it comes to creating trust it is not enough!

Subsequent research has demonstrated that all three factors are important in building trust and/or overcoming mistrust. For example, a leader may be viewed as highly competent, open and honest; however, if she/he is perceived as uncaring then trust cannot be built or sustained, and any existing mistrust will not be overcome.

Interestingly, in terms of *creating* trust, the *integrity factor* has emerged as the most significant, while the *care factor* (benevolence) has been found to be the most powerful component in terms of *overcoming mistrust* (Conchie et al., 2011).

Research by Carrillo (2020) echoes such conclusions. Moreover her study demonstrated the clear link between perceived care and safety performance. In describing her findings she stated that

Employee perception surveys showed that OSHA recordables were higher at sites where the senior manager was perceived as not caring about people. They also rose after a manager who was perceived as not caring replaced a respected leader who valued employees. I was further convinced of the connection between trust and performance when supervisors who received high trust scores on perception surveys ran accident-free, productive crews in facilities where supervisors with low trust ratings had high injury rates and poor production.

(pp. xi–xii)

The jury is still out as to why so many safety *thought leaders*, global safety consultancies, organizations and managers spend so much time advocating slogans, platitudes and outdated modalities and 'doing what the others are doing' instead of focusing on tangible factors that research has shown to be highly impactful on safety performance. Perhaps imposing yet more procedures, checklists, rules and systems is simply seen as easier than focusing on trust, care and psychological safety ('soft skills are hard!'). Surely though, leaders owe it to their teams to *do what works* rather than doing what is easy or comfortable.

Any organization serious about reducing, minimizing or even eliminating harm needs to first create the trust required for such lofty and laudable goals to be realized. More of the same simply won't do it!

Our safety, wellbeing and cultural development programs are evidence based, and we settled on the name *Care Factor* (given the research cited above, the reason for choosing that name is no doubt obvious).

An added bonus for our consultancy is that the *Care Factor Program* branding tends to result in the self-selection of clients who possess the maturity, humility and courage necessary to enable the program to reach its full potential, with a few would-be clients judging the key word (*care*) as too 'soft' for their workforce (and we know this has happened in a small number of cases).

Imagine that! A senior leader (or leadership team) decides that their workforce is not sufficiently 'grown up' to experience a change program with the word 'care' in the title. We know from experience that team members (as the program unfolds and the context becomes clear) really welcome the messages about trust and care and frequently make comments after the workshops such as "Brilliant, our leaders really need to hear this message too!"

All too often, managers project their own limiting beliefs onto their teams. Such occasions frequently point to a leadership team that lacks the emotional maturity to really get behind a trust-based program, so there would be little point moving forward without first doing some intensive coaching at a senior management level. The sad irony is that some of these very companies zealously emphasize their goal of 'Zero Harm,' and in such cases it becomes obvious to observers (and indeed the workforce) that the stated aspiration is a meaningless platitude more likely to create cynicism than any intrinsic motivation to buy into the goal (more on this topic in Chapter 2).

That such thinking still exists in a country (Australia) where a construction worker takes his/her own life every second day and around 200 people per year die as a result of workplace incidents is alarming and untenable. When senior leaders view the word 'care' as outside their (or their people's) comfort zone, it raises questions on how such apathetic organizations can possibly deal with any emerging mental health challenges among their workers.

So how did we get here?

Many organizations such as those described above are in traditionally male-dominated and *macho* industries. Moreover, young males in Australia (and other nations) tend to receive early childhood messages such as "big boys don't cry," "harden up," etc., so actually displaying vulnerability and demonstrating care can be daunting propositions for many leaders in such industries.

Nevertheless, leaders need to lead! It is time for the macho organizations to (psychologically) grow up and demonstrate the emotional maturity, integrity and care required to genuinely promote a psychologically safe workplace. We're leaders, not gang members!

Part 2 of this book will offer the reader strategies, skills and suggestions that enable leaders to create trust (and/or overcome mistrust) within their teams; however, an important first step on this journey from compliance to care is to understand where organizations are currently (and often inadvertently) undermining the psychological safety of their teams.

Some trust-wrecking behaviors are obvious (yelling, demonstrating a lack of integrity, name-calling, lying, etc.). However, what is less apparent to many is that some well-established (and usually well-intended) 'safety' practices often have the unwanted effect of increasing fear, mistrust and cynicism.

In the next two chapters we will put some of safety's 'sacred cows' on the grill, and at the end of each chapter you will be invited to reflect upon how you believe such approaches impact upon the trust and psychological safety of your own team/organization.

KEY POINTS

- Trust and psychological safety are strongly associated with more mature safety cultures and safety performance.
- Where fear is high, trust will be low and the resulting lack of psychological safety will impede the development of a 'speak-up,' proactive safety climate.
- Trust is experienced by team members based upon their perceptions about a leader's integrity, ability and benevolence (care).
- The care factor is the most powerful for overcoming team mistrust.

REFLECTION QUESTIONS

- Of the three factors associated with trust (integrity, ability and care), which do you believe you need to work on the most? Why do you come to that conclusion?
- How do you currently demonstrate care to your team?
- What actions could you and your leadership teams take to start building (or increasing the amount of) trust in your organization?

2

'ZERO HARM' AND OTHER PLATITUDES

Research shows that if goals are unrealistic, but you can achieve them by cheating, then people will cheat. They will commit fraud to obtain the incentive.

– George Loewenstein

Imagine there was a company that decided to go the extra mile in driving performance by implementing a goal of 'zero errors.' The ambitious target was made a 'core value' of the organization, and 'zero error' posters were proudly displayed around the offices (even in the bathrooms!).

Inevitably though, shortly after the grand launch, somebody made a mistake. Morale took a nosedive as investigations were conducted into how this could have happened. 'Zero error reset' workshops were rolled out, along with revised policies and procedures designed to prevent a repeat of the devastating occurrence.

Witnessing all the fuss, audits, and inquiries (and the associated mountain of paperwork), fear among staff members began to increase ("What if it was me that made the error?"). Due to basic human fallibility, others also made errors. Nothing huge or dramatic – just minor mistakes. As nobody else had witnessed these errors, the fearful employees made an understandable decision not to report them.

In the meantime, a blissfully ignorant management team was feeling proud of its '30 days error-free' status, and barbeques were held to celebrate the achievement, along with the apparently error-free staff members receiving celebratory caps, key rings, and other such trinkets.

Sadly, however, due to a lack of subsequent learning from these numerous small (yet unreported) mistakes, a very serious error occurred.

Sound familiar?

Of course, the above is a fable – pure fiction (at least I hope it is!). The whole notion of 'zero mistakes' is absurd, given the fundamental imperfection of human beings. Yet many organizations are seemingly in denial about the fact that adopting a goal of 'zero harm' often plays out in exactly the ways I described above.

I think in their finer moments most leaders would agree that, at some stage, an injury has been reclassified to avoid falling into the 'lost time injury' (LTI) category (i.e., someone has been injured but put on 'light duties' to avoid it counting as an LTI). It happens and it happens a lot! Think what that does to trust levels among employees (particularly when the reclassification is motivated purely by metrics).

RESEARCH INTO 'ZERO'

So does a 'zero harm' goal work? Well, probably not!

As Dekker (2017) noted, no studies reported in Zwetsloot et al. (2017) or elsewhere have been able to single out the presence or absence of a 'zero harm' vision (as a separate variable on a comparative basis) so as to determine its effect on safety outcomes. Hence there is no research demonstrating that 'zero harm' (as a clearly defined variable) reduces incident rates.

Moreover, zealous devotees of 'zero harm' would do well to temper their blind enthusiasm based on a recent study conducted in the UK construction sector which concluded that

> working on a project subject to a zero safety policy or programme actually appears to slightly increase the likelihood of having a serious life-changing accident or fatality; a possible 'zero paradox' … they suggest that the apparent trend towards abandoning zero amongst some large organizations is well-founded. As such, if zero policies stymie learning whilst failing to reduce accidents, the need for a countervailing discourse is clear.
>
> (Sherratt & Dainty, 2017)

And indeed, there *is* a countervailing discourse. In some cases the 'anti-zero' campaigners are just as fanatical as some of the more fervent 'zero harm' supporters. If you think I am barbequing this particular sacred cow, I suggest you pick up a copy of Dr. Robert Long's *For the Love of Zero* – he goes the full nuclear option! My own view is that such extreme positions are seldom helpful.

Most of the companies we work with have some form of 'zero' vision. This doesn't mean they are somehow stupid, unethical, immoral, or even (as some have opined) evil, and to suggest otherwise is ignorant and churlish.

I am a strong advocate of always looking for a positive intent – it's a powerful and underutilized leadership tool of influence. Rather than condemning a leader (or company) for engaging in seemingly errant or self-defeating choices (such as introducing a 'zero harm' goal), first look for the positive intent:

• What was the expected payoff (i.e., the positive intent)?
• What could the unintended negative consequences of that approach be?
• Could the payoff be achieved by other, more constructive means?

The 'zero' vision seems admirable and, I believe, is (generally) put in place by sincere people, with the aims being to focus attention on avoiding harm and making it clear that no harm is acceptable to them.

However, just because a goal has a positive intent it doesn't mean it's helpful.

Much has been written about the importance of goals being SMART, and while the acronym has numerous variations, the following captures its essence.

• Specific
• Measurable
• Achievable
• Relevant
• Time based

Even a modicum of analytical thinking will demonstrate that 'zero' goals struggle to fit the SMART criteria (or if they do they have to be shoehorned in!).

Is 'zero' specific? As a binary goal it would appear to be very specific indeed … but hang on a minute! What does harm include? First aid injuries? A stubbed toe? Does harm include psychological harm (e.g., stress leave, burnout, anxiety, depression, bullying, and suicides in mining camps)? If not, why not? Even in this brief analysis, what is meant by 'harm' becomes quite ambiguous – and if the goal is not explicit it becomes difficult if not impossible to measure.

Is it achievable? Again, that depends on how you have defined harm, and if what constitutes 'zero harm' is ill-defined, how will we know if we have achieved it? More to the point, *does your workforce believe the goal of 'zero harm' is achievable*? In my 20 years of experience I have rarely encountered a site where the majority of the workforce (or even the leaders) view 'zero harm' as an achievable goal. Does this matter? Absolutely!

When confronted with contentions such as those above, some of the more dyed-in-the-wool 'zero' cheerleaders will backtrack and say, "Oh well, it's an *aspirational* goal." To me that is like the perennial politician talking about achieving a budget surplus ... an aspirational goal. And as the voting public hears these words they roll their eyes and say, "yeah right!"

Maybe aspire to a better goal instead? A SMART goal? A goal that won't result in the workforce rolling their eyes?

Many organizations that fall into the category of proactive or integrated cultures – for example, high reliability organizations (HROs) – frequently achieve sustained low incident rates without so much as a mention of the word 'zero,' so clearly it is possible to achieve excellence without employing binary goals and running the risk of building scepticism and mistrust among the workforce.

In a recent (and excellent) podcast about 'zero harm' by Griffith University's safety science gurus Drew Rae and David Provan (2020), the researchers summarized the findings from all relevant and available studies. Their balanced conclusion was that *if a company doesn't already have a goal of 'zero,' it probably shouldn't adopt one*. They went on to say that if a company *does* have a 'zero' goal, leaders need to make it clear that it is in place primarily to direct attention toward safe operations, rather than the numerical goal *per se*.

I largely concur with David and Drew; however, for the companies that retain the 'zero' goal, I would add, "stop banging on about it!" There's a good chance that every time leaders mention it, somewhere in the room many of their workers are rolling their eyes!

As well as 'zero harm,' another often-used phrase in organizations is "Your safety is our highest priority." If that is true – if the phrase is to be anything other than a mere platitude, then senior staff need to start leading by means of authentic communication rather than glib, patronizing slogans that the workforce hears as hollow rhetoric. Leaders would do well to follow the examples of more mature organizations by doing what has been shown to be valid and evidence based, rather than what is easy or comfortable.

KEY POINTS

- Binary safety goals (e.g., 'zero harm') are not supported by the research; in fact 'zero' programs have been found to be associated with increased occurrences of serious incidents and fatalities.
- Most employees view 'zero harm' as unrealistic and unachievable; hence the goal is likely to be regarded as a mere platitude resulting in cynicism and mistrust.
- When companies become overzealous about a 'zero' goal they can become intolerant of incidents, and their people may become less inclined to speak up due to fear of reprisals and a lack of psychological safety.

REFLECTION QUESTIONS

- Is your company's stated safety goal likely to be increasing or decreasing trust? Do you buy into it? Does the workforce?
- What is the intent of the goal? Has this been made clear to the workforce?
- What conversations could be had (and with whom) that could result in a move away from hollow rhetoric and safety platitudes?

3

BBS (AKA BEHAVIORAL BULL Sh#t!)

Behaviors are not the problem – they are expressions of the problem.

– Clive F. Lloyd

We are entitled to our opinions – we are not entitled to our own facts!

Lately it seems empirical evidence, peer-reviewed research, and being guided by experts have ceded authority to the exposition of baseless opinions and (to use the current parlance) 'fake news.'

While evidence can (and does) provide sound policy direction on matters of great importance such as vaccinating children against measles and polio, increasingly these days populist politicians seem enamored by the unsubstantiated ramblings of vested interest groups, conspiracy theorists, and common-or-garden nutjobs! These non-experts seem to be extremely adept at gaining social media exposure, through which channels they find an all-too-willing audience to indoctrinate with their self-serving version of 'the truth.'

Favoring opinion over evidence has also been commonplace in the safety arena for decades, most notably in the continued use of (and advocacy for) behavior-based safety (BBS) – or as Phil La Duke once labeled it, 'blame-based safety'!

I don't know anyone in my large network of psychologists who would use behaviorism as a modality of choice (unless they are training their dogs of course). Yet, despite the fact that the approach fell out of favor decades ago, BBS (which is based on the underlying principles of behaviorism) is still distressingly commonplace in the mining, oil, and gas and construction sectors today.

BBS has made a positive difference, just as leeches have historically been useful in some medical procedures. That doesn't mean I want bloodsucking parasites to be my doctor's 'go to' diagnostic tool or treatment method! (Give me the antibiotics please!)

I believe BBS may have been useful for companies inhabiting the lower realms of cultural maturity. That is, the approach may have assisted apathetic/reactive cultures to make their way toward a more 'involving' climate by at least increasing the focus on safety. However, as many companies have found, to their chagrin, moving toward proactive/integrated cultures using BBS is extremely challenging.

This is largely because, at a fundamental level, behaviorists believe that consequences are the motivating force behind behaviors – that is, positive reinforcement rewards a person for a 'good' behavior, whereas negative reinforcement results in unwanted or unpleasant consequences (punishment). If understanding human behavior was truly this straightforward, all leaders would need is a stick and a bunch of carrots!

At best, behavioral approaches based on external control may create short-term compliance – they do not create commitment – and as soon as the external threat/reward is relaxed, so too is the level of compliance. Moreover, most people are, after a while, very likely to breach safety rules because our brains are actually hard-wired to do so. That is, we (consciously or unconsciously) seek to modify routine tasks to make them quicker or easier – this is what complacency is about – and we all do it! Hence, safety approaches relying solely on the basis of compliance will generally fail in the long run (Read et al., 2010).

The essential message here is that people are not robots (or, indeed, dogs or pigeons!), and one size does not fit all. Managers may errantly believe that if a specified antecedent is present it will result in a particular behavior – but the subsequent behavior is likely to be different from one individual to the next. An individual's response will be unique, and based upon his or her own internal processing, the individual controls this, not the stick-wielding manager (Read et al., 2010).

Finally, the workforce often perceives behavior-based safety programs in a negative way. BBS can be seen to be focusing on failures, catching people doing something 'wrong' (often evident in *Safety Walks* and *Safety Observations*), or 'ratting out your mate.' They can (and often do) result in blaming the worker involved as being the root cause of the problem (common in apathetic and reactive cultures), which in turn creates mistrust and inhibits future reporting as illustrated in Figure 3.1.

DEEPWATER HORIZON INCIDENT

On April 20, 2010, BP's now-infamous Deepwater Horizon rig blew up. The explosion killed 11 workers and resulted in the release of

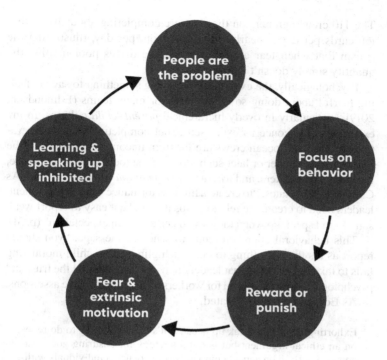

Figure 3.1 The fear loop.

5 million barrels of oil into the Gulf of Mexico from the Macondo well deep below the ocean floor. Subsequent investigations concluded that the disaster was a result of a series of decisions and actions that increased risks, yet failed to adequately analyze or mitigate them.

BP (who incidentally were – and are – proponents of a 'zero harm' goal) and operator Transocean also used traditional BBS tools such as *Safety Observations*. The Transocean policy for such observations included the completion of a daily START card. During the post-incident inquiries, crew members shared their feedback about the process and proffered comments such as:

- " … the focus was on quantity not the quality of observations."
- "People [tried] not to rat people out (so to speak), you know, like you wanted to be helpful, whereas some of the higher-ups in the office, they kind of wanted to weed out problems."
- "I've seen guys get fired for someone [writing] a bad START card about them." (pp. 143–144, Vol. 3, CSB Macondo Report).

The 110 crew members on the rig were completing about 100 safety job cards per day – nearly one per person, per day, illustrating the truism that when fear around a procedure drives poor quality, the quantity simply doesn't matter!

Psychologically, the experience of having something to say yet feeling fearful about doing so is the norm for many teams (Edmondson, 2019), particularly in overly hierarchical *parent–child* cultures. In my experience, the concerns expressed about completing safety observations by the Transocean crews are far from uncommon. Even with the best of intent, sooner or later such tools put the focus (and by inference, the blame) on workers, and *where there is fear you will get bad data*. As Carrillo (2020) states, "to create a high performance work environment, leaders need to create the relationships that make it easy to report weak signals and speak up when they are uncertain of an expectation" (p. 31).

This behavioral approach tends to send the message, "you should report as it's the right thing to do." Such glib, parent–child moralizing fails to take into account that leaders may not have created the trust and psychological safety required for workers to act on such naive assertions. As Edmondson (2019) stated,

> Exhorting people to speak up because it's the right thing to do relies on an ethical argument but is not a strategy for ensuring good outcomes. Insisting on acts of courage puts the onus on individuals without creating the conditions where the expectation is likely to be met.
>
> (p. 82)

Behaviorism (and by extension BBS tools such as those described above) tends to ignore internal processing and focuses instead on a basic stimulus–response approach.

REWARDING 'GOOD' BEHAVIOR

A core tenet of behaviorism is *positive reinforcement* – providing rewards for positive behaviors. This can work well with dogs – just give Rover a treat when he brings the stick back, and he'll keep doing it. Humans are a lot more complex.

Unfortunately, many organizations still employ the risky BBS practice of providing a 'safety incentive' (e.g., a cash bonus) for 'good' behaviors (e.g., no injuries during a set reporting period). Just don't! Such a primitive and patronizing behavioral approach can backfire with catastrophic consequences.

An aluminum smelter owned and operated by a global mining company had a policy similar to the above, with significant cash bonuses paid for remaining free of lost time injuries (LTIs) during a set period of time.

With just days to go until the end of the reporting period, a forklift operator (who was not wearing his gloves) lacerated his thumb while trying to remove a jammed pallet. A colleague (who was keen to suggest that they keep the incident quiet) assisted the injured operator in dressing the wound and bandaging it, and the operator then hid the bandage by keeping his gloves on. The incident was not reported, and the operator did not seek medical attention, as this would have alerted management to the significant injury. A few days later the reporting period was completed, and all staff members had bonuses deposited into their bank accounts.

With the cash in the bank, the injured operator (who had been in great pain since the incident) went to his doctor, only to be told the wound was so badly infected that he needed to go to hospital immediately. His thumb had to be amputated due to the severe level of infection and other complications.

The truth came out, everyone was made to return his or her bonus, and the forklift operator was fired along with his 'helpful' colleague. Bonuses for lag indicators (such as LTIs) are a bad idea as they invariably lead to non-reporting. Understanding human motivation requires us to look beyond the simple stimulus–response models espoused by radical behaviorism.

As Smith (1999) stated,

> There were many other things that psychologists needed to explore such as motivation, perception, creativity, problem-solving, experience, and interpersonal relations. Eventually, new data was gathered on these subjects and raised questions that behaviorism couldn't explain. This brought about a paradigm shift and led the way to a new theory of psychology – known as "cognitive science." The field of psychology may have advanced beyond behaviorism but in the field of management it is quite a different story. Behaviorism is still applied with a vengeance by managers. The fact is, it is very useful in a command and control management system. We have applied it with such force and magnitude we now accept its premise without question.

Despite the above being written 20 years ago, and having been followed by a plethora of similar critiques, BBS is still promoted by many, especially purveyors of BBS programs (speaking of blood-sucking parasites!).

Arguably, widely disseminated 'fake news' may not be of any great significance when it is about celebrity weddings or other trivial matters. However, when we are talking about the safety of our people we cannot afford to simply assume that using a common approach is justified merely because it has been around for decades and 'everyone else is using it.' Most thinking people still vaccinate their children against measles and polio, despite the ever-growing chorus of conspiracy theories insisting that immunizing children causes autism (it doesn't!). If we are capable of such rational decision making about the health of our children, surely as leaders we are also able to make informed decisions about evidence-based practice when it comes to safety.

KEY POINTS

- BBS relies on a stimulus–response, reward-or-punish approach that enables (or embeds) a parent–child hierarchy and a 'blame the worker' mentality.
- It is not an approach that lends itself to creating the trust and psychological safety required to move toward a more mature safety culture.
- Providing incentives based on lag indicators can lead to under reporting.

REFLECTION QUESTIONS

- Does your company provide rewards based on lag indicators?
- In what ways is your current BBS program working well?
- What concerns do you have about your BBS Program?
- On balance, do you believe your current BBS program will assist your company to move toward the more mature levels of safety culture?
- If not, what conversations could be had (and with whom) that could begin the process of improving or moving beyond your current BBS program?

PART 2

4

ORGANIZATIONAL VALUES
OR COMPANY PLATITUDES?

It's not hard to make decisions when you know what your values are.

– Roy Disney

Not so long ago there was a saying among frequent flyers: "If it 'ain't Boeing, I 'ain't going," such was the level of confidence about the safety and reliability of Boeing's aircraft – how times change! In October 2018, Lion Air flight 610 crashed into the ocean off the Indonesian coast. Five months later, Ethiopian flight 302 met a similar fate. In total, 346 people died.

The crisis surrounding Boeing's 737 MAX Aircraft is the most severe the aviation behemoth has faced in its century of operation. As well as the financial losses (which have exceeded $US28 billion), the damage to the company's once-impeccable reputation is inestimable.

As former Boeing Senior Manager Ed Pierson stated, "Something happened in the translation from, 'let's build a high-quality, safe product,' to 'let's get it out on time'."

Boeing's core values are:

- Integrity
- Quality
- Safety
- Diversity and inclusion
- Trust and respect
- Corporate citizenship
- Stakeholder success

The above illustrates the profound differences between *espoused* and *in-use* values. A company's espoused values are *what they say* its values are, whereas the in-use values are those reflected in its *actual ways of operating*. Clearly, at least in terms of the 737 MAX project, Boeing's senior leadership decisions were not aligned with the organization's core values.

I work a lot with companies that are seeking to develop their culture, and a fundamental starting point in our approach is to ask some key questions about values, such as:

- What are your company's core values? (You may be surprised how few senior leaders even know what they are!)
- Are these espoused or in-use values?
- How do you think your employees, clients, and stakeholders would answer the above question?

Gaining clarity around these questions can assist an organization to understand whether their values are contributing positively to the culture, or actually undermining it. If company values are well known, understood and used as decision-making tools by leaders, employees are presented with clear evidence that they can trust their management team. Conversely, if the organization's values are merely words that look good on posters or the company's website but are barely known, let alone utilized, then there is a heightened risk that leadership decisions will be made based on political expediency and short-term payoffs. The workforce will always notice such inconsistences and generally conclude that management is hypocritical and cannot be trusted.

I've never seen values on a company's website that came with an 'escape clause' – that is, "We behave with integrity, unless ..." – yet this is precisely how some companies operate, only to subsequently face the inevitable losses in terms of reputation, trust of clients and the psychological safety of their teams.

As leaders, our own personal values represent how we see ourselves at our best – *our authentic selves*. Bill George, the Godfather of the *Authentic Leadership* approach, posits the question, "what guides you when you reach a fork in the road?" At such times we can be tempted to stray from our values, perhaps feeling a need to wear a psychological mask to fit in, to gain approval or to emotionally protect ourselves.

We also encounter such dilemmas in our professional lives. As leaders, we may acknowledge our company's values make one road obvious; however, if we took the other road there could be significant payoffs: it may be quicker, cheaper, easier, my boss may like it, I may get kudos, etc.

Authentic leaders are values-based, and they use their values as *decision-making tools*. Put simply, if an alternative road is not aligned with their values, they simply won't go down it, regardless of any short-term gain. Leaders who consistently allow their values to guide choices build trust within their teams. As Bill George (2003) states, "Authentic leaders demonstrate high integrity by taking courses of action independent of pleasing their audience or political expediency. They walk their talk, and in doing so create a level of trust that allows others to follow."

Any organization serious about wanting to build trust and psychological safety would do well to start with an authentic appraisal of its stated core values, as well as the degree to which they are known and appropriately utilized. Any leader seeking to do the same with his or her team would benefit from similar reflections upon his or her own personal values.

How dearly would Boeing's senior leaders love to go back in time and stick to their core values?

AUTHENTICALLY DEMONSTRATING CARE

Authentic leadership is the full expression of "me" for the benefit of "we"

– Henna Inam

Whenever I have a group of leaders together, I run the following brief activity – try it with your leadership group and let me know what insights you gained.

1. I ask group members to raise their hands if they believe that, generally speaking, leaders *do care about their teams* – most (if not all) of the hands will go up.
2. Next I ask the group members to raise their hands if they think that, in general, leaders are *good at demonstrating care for their people* – far fewer hands go up.

In my experience, most leaders do in fact care about their teams (though it is far from unanimous). It's not, therefore, that leaders *don't care* – it's more than that, either they feel uncomfortable displaying it or they aren't particularly good at doing so. This then begs another question – if leaders are not expressing it, how would their teams know their leaders *do* care?

Unsurprisingly then, when I run the same exercise at the team level and ask team members to raise their hands (if they believe their leaders care about them), far fewer hands go up.

As Conchie (2013) stated,

> leaders' interactions with employees – what might be termed their leadership style – will play an important role in the establishment of trust relationships. Evidence is mounting that a leadership style, where leaders develop affective bonds with their employees will help facilitate trust development and positively influence safety.

I witnessed a great example of safety leadership (and just plain leadership) a little while ago, which elegantly highlighted the benefits of developing the 'affective bonds' that the researchers refer to.

NEXT GENERATION SAFETY LEADER PROFILE 1: JOHN PHILLIPS

John Phillips is the HSE Manager with the University of Melbourne's Project Services team, which is essentially the university's construction division. I have known and respected John's leadership for many years, and he always opens and closes when we facilitate *Care Factor* programs with his teams. ('Nigel' could learn a lot simply by observing John!)

While taking a coffee break after opening one of our workshops, John started chatting with one of his construction crews. He noticed a new team member – a young guy – who was also on a break and went over to introduce himself. John asked the new guy what his role was on the team, and he responded, "I'm just the spotter." Picking up on the word 'just,' the Sage Safety Manager asked him about what that role entailed (keeping people out of danger, etc.), and the discussion then moved to the importance of the role and the inherent trust placed in him by his team. At the end of the chat the young guy was puffed up and highly motivated to put his fresh eyes to excellent use by actively seeking out areas for improvement that he could share with his team and John during their next chat.

No KPIs, no planned 'safety walk' – just a great leadership conversation opportunity that was brilliantly taken. It's not dramatic, flashy, complex, or technical. There was no paperwork involved – just authentic, relationship-based leadership that demonstrates care and builds trust – and it leaves everything else in its wake!

I posted the above observation on LinkedIn, and at last count it had received over 56,000 views and 724 'likes.'

I think John himself would be amazed at the sheer volume of responses, let alone the admiration voiced in the comments. John was simply being John, authentically. I also found myself a little surprised that the simple social media post gained such high levels of engagement. It suggests that John's approach really resonated with people. Many commented on the excellence of the interaction and lamented that this style of leadership communication is so rare. Indeed it is, and in a moment of frustration, I found myself wondering why – if so many people recognize John's approach as a brilliant example of safety leadership – so few do it?

While it seems like a simple conversation, let's just break down what actually happened.

1. John was *alert to opportunities* to build relationships and trust.
2. He *noticed* there was a new person on the team.
3. He *authentically engaged* with the employee.
4. He *asked* a great question.
5. He *really listened* to the response.
6. He noticed an opportunity to *build purpose, esteem, and psychological safety*.
7. He assisted the employee to *reframe* a limiting belief.
8. He displayed genuine *care*.

In that brief interaction, John helped the new team member feel a sense of pride and belonging, increased his trust in leadership, and made it more likely the young guy would speak up in the future.

John's approach epitomizes visible, felt leadership, and no KPI-driven procedure can compete with its positive impact. As the splendid Rosa Antonia Carrillo says, "a policy can't make it safe to speak up." Yet a conversation like the one described above can go a long way toward building the psychological safety required for a 'speak up' culture to flourish. The mindset driving John's leadership style is illustrated in Figure 4.1:

Of course, John wasn't working from any particular leadership theory – he was just being who he is. Nevertheless, such authentic leadership – what some would describe as the effective application of *soft skills* – is supported by hard evidence as an enabler of trust and safety performance.

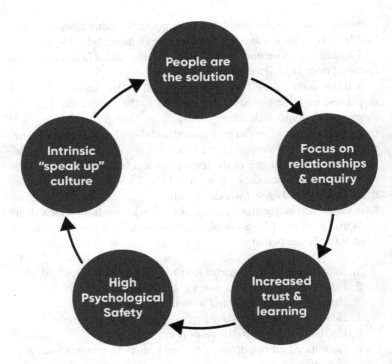

Figure 4.1   The trust loop.

As I mentioned earlier, focusing directly on the culture is not the best way to facilitate culture change – cultures are best developed one authentic conversation at a time. John's brilliant exchange with the young spotter is an example of such an interaction, and when this communication style becomes the norm, the *team culture* develops beyond the traditional parent–child dynamic. When authentic leadership becomes the norm across all work teams, the *organization's culture develops*. As Covey (2006) stated, "establishing trust with the one establishes trust with the many" (p. 25).

While reviewing the literature on safety leadership, one finding became crystal clear: The 'command–control' approach of the past is the antithesis of what is known to create proactive safety cultures. If such top-down hierarchical approaches are not dead, they are certainly coughing up blood!

There are, however, still pockets of resistance to this inevitable shift. At the end of a recent two-day leadership program at a mine site, a superintendent, 'Bill,' pulled me aside and said, "Thanks, that was great. But I'm 'old school,' I do wield the big stick and they sure know who's boss!" He went on to say that "You're working with my team tomorrow, and when you speak with them you'll notice two things: They don't like me, *but they bloody well respect me!*"

I did meet Bill's team the following day, and he was half right. They didn't like him; however, neither did they respect him – they feared him! Unsurprisingly, Bill's team members admitted that they rarely reported minor incidents or 'near misses' to Bill, as the consequences for doing so had historically been severe and sustained. It was not a shock, therefore, when I discovered that Bill's team had the highest lost time injury frequency rate (LTIFR) on site.

O'Dea and Flin (2001) found that more directive leaders (such as Bill) *tend to overestimate their abilities to motivate and influence the workforce.* Such archaic parent–child approaches simply cannot create the levels of trust necessary for people to freely report safety concerns or for teams to move toward more proactive or integrated safety climates.

The limitations associated with BBS methodologies and traditional top-down management approaches have provided the impetus for the next generation of safety leaders to explore more humanistic, values-based models such as authentic leadership.

The concept of authentic leadership comes from the positive psychology movement and focuses on the positive attributes people have, rather than dwelling on what is (assumed to be) 'wrong' with them. The basic thesis of the approach posits that people are inherently motivated and seek commitment, responsibility, and enjoyment from their work. As such, authentic leadership shares a great deal of overlap with the current philosophies espoused by advocates of the *Safety Differently* movement.

While research efficacy on *Safety Differently* is currently sparse (no doubt due to its relatively recent emergence), given that the approach's core principles mirror those of authentic leadership (e.g., engaging the workforce, focusing on positive attributes), I believe it holds great promise as a vehicle for safety culture development, and will be a focus of the final chapter.

I base such optimism on the fact that authentic leadership is amassing a significant amount of research support in the safety literature (e.g., Avolio et al., 2004; Dirik & Seren, 2017; Mearns, 2008; Cavazotte et al., 2013; Nielsen et al., 2013). Such studies have consistently demonstrated that the approach significantly predicts a positive safety climate, and have also confirmed a link between authentic leadership and safety performance. Indeed, researchers concluded that organizations would do well to consider recruiting and developing authentic leaders to foster positive safety climates and risk management (Nielsen et al., 2013).

Does your leadership style more resemble John's conversation with the young spotter, or Bill's 'old school' leadership philosophies? The degree of trust (or mistrust) within your team will be reflected accordingly. Leading authentically based on your own core values and consistently demonstrating care to your team can be express routes to the creation of psychological safety.

KEY POINTS

- Straying from core values can lead to catastrophic decisions, particularly when organizations face time or financial pressures.
- Values can be merely espoused or in-use guides to consistent and ethical decision-making.
- Authentic leaders focus on positive attributes of their people and adopt a collaborative, adult–adult leadership style.
- Values-based, authentic leaders consistently demonstrate care for their people, creating trust and psychological safety.

REFLECTION QUESTIONS

- What guides you when you reach a fork in the road? What are your own personal values?
- Do your people know your company's core values?
- Are the ways your company operationalizes its values likely to be increasing or decreasing trust?
- How does your leadership team demonstrate care for the workforce?
- How do you rate yourself and your fellow leaders in terms of authenticity?
- What needs to change?

5

WHAT'S YOUR STORY?
Building Trust through Self-Disclosure

> Vulnerability is the birthplace of innovation, creativity and change.
>
> – Dr. Brené Brown

I remember meeting 'Phil' for the first time, as well as the instant dislike I took to him. I had just joined the clinical team in a drug and alcohol rehabilitation facility, and Phil was one of the therapists. He seemed sneering and angry, constantly sarcastic to peers and patients, yet sucked up to the management and seemed to have an endless stream of stories to relay to them about his alleged brilliance.

My immediate aversion to Phil was intensified and generalized by the 'halo effect' (the tendency for an impression created in one area to influence opinion in another area), and very soon he could do nothing right in my eyes.

Of course, my negative judgments about Phil rapidly became self-fulfilling prophecies, and in an alarmingly short period of time, he and I were members of a mutual loathing society. This unhelpful relationship impacted negatively upon my work, the organization's culture, and – worst of all – the patients.

Then something very simple but profound occurred. As part of the clinical team's group supervision process, we were all invited to share our stories – our life journeys to that point. Phil shared incredibly authentically and courageously about the abuse and neglect he experienced throughout his formative years – how his older siblings were treated differently, and how his parents seemed aloof and dismissive of any of his achievements.

Phil's sharing enabled me to make sense of some of his current behaviors. He was still seeking approval from authority figures and was threatened by his peers, so he was often dismissive or hostile toward them. After his sharing, it was easy for me to approach Phil and to thank him for his courage and honesty. While Phil and I never became great friends, my newfound empathy enabled me to move beyond my own reactivity and meet him within a working relationship of mutual respect and understanding.

We all have a story. We have walked through different parks and knelt at different graves, and our life experiences have shaped who we are. Few of us had perfect childhoods, and where our fundamental needs were not met, we adapted our behaviors through various protective coping mechanisms such as wearing psychological masks.

Our lazy brains' preference for System 1 thinking (subconscious, fast, automatic, frequent, emotional, stereotypic) over System 2 thinking (conscious, slow, effortful, infrequent, logical, calculating) means that we generally form opinions about people very quickly, particularly when we don't know their background story.

Steven Covey illustrates this point brilliantly in his classic work, *The 7 Habits of Highly Effective People.*

I remember a mini-Paradigm Shift I experienced one Sunday morning on a subway in New York. People were sitting quietly – some reading newspapers, some lost in thought, some resting with their eyes closed. It was a calm, peaceful scene. Then suddenly, a man and his children entered the subway car. The children were so loud and rambunctious that instantly the whole climate changed.

The man sat down next to me and closed his eyes, apparently oblivious to the situation. The children were yelling back and forth, throwing things, even grabbing people's papers. It was very disturbing. And yet, the man sitting next to me did nothing.

It was difficult not to feel irritated. I could not believe that he could be so insensitive to let his children run wild like that and do nothing about it, taking no responsibility at all. It was easy to see that everyone else on the subway felt irritated, too. So finally, with what I felt was unusual patience and restraint, I turned to him and said, "Sir, your children are really disturbing a lot of people. I wonder if you couldn't control them a little more?"

The man lifted his gaze as if to come to a consciousness of the situation for the first time and said softly, "Oh, you're right. I guess I should do something about it. We just came from the hospital where their mother died about an hour ago. I don't know what to think, and I guess they don't know how to handle it either."

Can you imagine what I felt at that moment? My paradigm shifted. Suddenly I saw things differently, I felt differently, I behaved differently. My irritation vanished. I didn't have to worry about controlling my attitude or my behavior; my heart was filled with the man's pain. Feelings of sympathy and compassion flowed freely. "Your wife just died? Oh, I'm so sorry. Can you tell me about it? What can I do to help?" Everything changed in an instant.

While we remain aloof and keep our emotional distance from others, it is all too easy to judge and condemn those who appear different from us. It seems to me of late that political expediency is often served by focusing on the (often confected) differences between groups of people, with a toxic and self-serving goal of amplifying fear and disdain of 'them' by 'us.' History is replete with dictators who employed inflammatory orations to tap into people's inherent prejudices for their own political or ideological gain. Understanding someone's story can provide us with the empathy and compassion required to see through such noxious rhetoric and to break down unhelpful assumptions and stereotypes.

Technology has enabled us to be more connected to our workmates than ever before. Increasingly though, emails and text messages are replacing face-to-face communication, resulting in diminished opportunities to build social and emotional bonds. Unsurprisingly then, trust in organizations and business leaders is declining, along with an associated reduction in staff engagement and an erosion of company culture. Leaders who take the time to share their stories (and listen to others') are helping to reverse this trend and build psychological safety within their teams.

It's unlikely we will like or get on with all of our work colleagues, or everyone in our leadership team, yet by making a commitment to authentically connect with our teams, and better understand each other's journey, we can create more trust and empathy and a more inclusive, caring culture.

To a degree we all wear 'psychological masks' at work to fit in, to reduce anxiety, and to feel comfortable. These masks both shape and are shaped by the culture we work within, and seldom in a positive way. For example, a 'tough guy' mask can inhibit self and others from admitting mistakes, reporting incidents, or speaking up when experiencing mental health challenges.

A colleague of mine was recently facilitating a workshop on stress management with a group of senior leaders in the oil and gas industry. He began by asking the group to discuss (in pairs) the indicators (emotional, psychological, physiological, behavioral, etc.) that help them recognize when they are experiencing stress. After a while he noticed one pair of leaders – the most senior in the room – sitting in silence. He asked how they were doing, and one of the leaders responded, "Y'know, we were just saying that we've spent our entire careers in the oil and gas industry shutting down these symptoms ... to the point that we're now struggling to identify them!"

Suppressing emotions, moving into denial, and adopting psychological masks are all too common in the macho industries, and this move away from authenticity is damaging not only to the trust levels in our teams, but can also negatively impact our own psychological and physical wellbeing.

What could be different if our team members felt psychologically safe enough to lower their masks? Authentic leaders create trust by dropping their own masks and, in so doing, make it just a little easier for their people to do the same. As Rosa Antonia Carrillo (2020) points out, inviting our teams to share their personal stories increases psychological safety. While observing team members who were sharing their stories, Rosa noted an interesting dynamic whereby *each story became more personal as the exercise progressed, as if the willingness of one person to be vulnerable opened the way for the next person to take more risks* (p. 35). This willingness to be vulnerable – to put yourself out there – is a courageous act, but the payoffs from leading such a powerful, authentic process can be immense.

GETTING STARTED

If sharing your story seems like a stretch target at present, here's another vehicle for self-disclosure that could feel less intimidating and act as a stepping-stone. As mentioned in the preface, your responses to the *reflection questions* at the end of each chapter will form a useful summary. By sharing the areas you are seeking to improve with your team, you will be displaying authenticity, humility, and a preparedness to be vulnerable. Moreover, disclosing your responses to your leadership team could be a powerful catalyst for wider discussions and subsequent change.

KEY POINTS

- Being prepared to 'drop our masks' and share our stories can be an express route to creating psychological safety within our teams.
- Our preparedness to be vulnerable helps others to share their own stories, increasing emotional connectedness and trust.

REFLECTION QUESTIONS

- How well do you know your team members? How well do they know your story?
- Are you willing to demonstrate vulnerability by sharing your story with your team? If not, what are some other ways you could demonstrate a degree of self-disclosure to really connect with your team?
- What do you see as the benefits of committing to the above?
- When will you start?

6

MIND YOUR LANGUAGE!

Help people think better – Don't tell them what to do.

– David Rock

Language is powerful. Our words are important. Few things are likely to have a more frequent or profound impact on the trust (or mistrust) levels of our teams than the words we speak on a daily basis.

Our words can influence our teams to frame events in positive or negative, helpful or hindering, and trusting or fearful ways.

As a simple illustration, a word I have chosen to remove from my vocabulary is 'should' (although it still slips out from time to time!). It is an example of a word that can be detrimental to relationships and trust and can usually be replaced by the word 'could.'

For example, contrast these two sentences:

1. "You should exercise more."
2. "You could exercise more."

In the first sentence the word 'should' can engender guilt if I decide not to exercise more, whereas in the second sentence, 'could' implies choice. In transactional analysis terms, 'should' is an example of a word associated with the 'parent' ego state. As such, it is likely to be received by the listener from his/her 'child' ego state. Other phrases that can have a similar impact tend to begin with words like:

- You must …
- You have to …
- You ought to …

Teams working in apathetic and reactive cultures will frequently hear these guilt-laden phrases from their bosses. The overuse of such language can embed parent–child relationships between leaders and their teams, stifling ownership, responsibility, innovation, creativity, and psychological safety.

The safety field has plenty of examples of hierarchical ('parent') terminology.

For example:

• Accident investigation
• Safety officer
• Safety audit
• Compliance officer

Who doesn't love being audited and investigated by an officer? Team members often lament the fact that safety managers treat them like children, and, given the prevalence of the above 'safety' language, it's not difficult to understand how they reach that conclusion – it doesn't have to be this way.

TRUST TALK

In a recent paper, Simon Bown (Head of Health, Safety, and Environment) described London Luton Airport's shift from using hierarchical 'parent–child' terminology to more inclusive, 'adult–adult' trust-building language.

> Plenty of impact has been made just by the pure use of language, as certain words carry stigma. Here are some of the changes: Advisors to Coaches; Accident to Learning Event; Accident Investigation to Learning Review; and Audits to Continuous Improvement Opportunities.
>
> (Bown, 2019)

This purposeful shift to consciously framing safety events, roles, and processes with less hierarchical *parent–child* terminology results in more trusting, engaged teams. As Edmondson (2019) noted, changing our language in these ways helps people "shift from a belief that incompetence (rather than system complexity) was to blame. This shift in perspective can prove essential to helping people feel safe speaking up about the problems, mistakes and risks they see" (p. 155).

'DIFFICULT' CONVERSATIONS

Perhaps there is no greater risk to psychological safety as there is when we encounter potentially difficult conversations. These may involve admitting a mistake, giving performance feedback, or simply expressing a contrary viewpoint. As leaders, it is vital we are mindful of our language on such occasions.

The following are some suggestions on how leaders can approach difficult conversations authentically, respectfully, and with a focus on positive outcomes and maintaining trust.

1. *Do a feelings check ... first!*

What makes a conversation difficult? If you have framed an imminent discussion as 'difficult,' it's likely that you anticipate elevated emotions within yourself, the other party, or both. I have heard some people suggest that they are reluctant to have a conversation because it may make the other person feel uncomfortable. However, in reality it's often less about the other person and more about how *we* would feel with the other's imagined discomfort.

Before we have the conversation it's wise to assess our own emotional state. If I'm angry, quite simply, now is not the time to have the conversation. Anger is just a feeling, one we all have from time to time, and there is nothing intrinsically 'bad' about that. However, when we project our anger onto others (an act of aggression), they can quickly lose trust and will likely become defensive (or match the aggression).

In these situations we need to remember the anger is ours – nobody else wants it. When we are angry, adrenalin and cortisol flow through our veins and drive a 'fight or flight' response. In this heightened emotional state we often raise our voices and our words are literally meant to hurt! Before you have the conversation, step away and allow time for the anger to subside. Consciously pre-framing the objectives of the conversation can help with this process, as illustrated below. Remember to:

- *Focus on the positive intent rather than the person.* When we're experiencing anger or frustration – say after witnessing what appears to be a breach of safety rules – often we can move into judgment and inference ("they are careless, reckless, thoughtless," etc.). In so doing, we can miss the positive intent that may have been driving the person's choices.

- By focusing on the positive intent we may just learn something about the differences between how *we imagine* work is done and how the work *is actually done* by the teams. At the very least we can now acknowledge the positive intent and also take the opportunity to ask about any associated risks (a learning conversation). If we instead focus on the assumed negative characteristics of the 'offender,' all we'll do is create fear and mistrust and any potential learning opportunities are lost.

2. *Place and time*

If the discussion is not part of an official process, then avoid formality. An informal conversation in a neutral venue will tend to lessen the chances of the other party adopting a defensive stance. Equally, if you anticipate the conversation is likely to be sensitive or potentially difficult, then ensure you don't initiate it in public (e.g., in a work area within sight or earshot of peers). In terms of timing, seek to have the conversation as soon as possible after the initiating event (i.e., the reason you are wanting to have the conversation). It's unwise to raise events that happened some time ago, as people understandably become curious about why it wasn't raised at the time – they may also wonder what else you are keeping up your sleeve – a real trust killer!

3. *Own your words*

Consider the following statements:

- "This is how it is …"
- "This should be common sense …"
- "You should …"
- "Why can't you just …?"

All of the above *parent–child* phrases leave little room for the other to have a different perspective without being made 'wrong.' The result is people shut down, become defensive, or just nod their heads, hoping for a quick end to a painful interaction.

Take ownership of the words you use. For example:

- "In my opinion …"
- "The way I see it is …"
- "To me, a possible way forward could be …"

In each case the speaker is not stating anything as a 'fact,' or inferring his/her position is the only perspective that can be taken – he/she is simply *owning his/her words and opinions*. This ownership permits the other to have a different perspective and still safely participate in the conversation without being made wrong.

4. *Invite others in*

Ownership phrases such as the above can then be followed by questions that invite the other's point of view, such as:

- "How do you see it?"
- "What's your take on this?"
- "What other ways forward do you see?"

We now have a two-way conversation where others are free to safely share their perspectives. Rather than driving fear and defensiveness, these approaches help maintain trust and psychological safety. Inviting our people to share their thoughts, ideas, and challenges is fundamental to creating psychological safety. This is often done through the effective use of questions and will be the focus of the next chapter.

KEY POINTS

- Language is powerful. As a leader, your words strongly influence the degree to which your people experience trust or fear.
- Hierarchical 'parent–child' language can diminish psychological safety within our teams.
- Monitoring our own emotional state, owning our words, and inviting our people in can help us engage in difficult conversations without others losing trust.

REFLECTION QUESTIONS

- To what degree do you use 'parent' language with your teams?
- Does your company use hierarchical language to describe safety functions and roles (e.g., accident investigation, safety officer, safety audit)? What changes could be made?
- What changes could you make to your communication style and use of language to foster psychological safety among your team?

7

YOUR TEAM HAS THE ANSWERS, DO YOU HAVE THE QUESTIONS?

Judge a man by his questions rather than by his answers.

– Voltaire

I strongly believe that the workforce has many of the answers that leaders seek, yet, in my experience, the 'wisdom of the crowd' is rarely accessed.

This is a waste of a wonderful resource and a missed opportunity to build trust, engagement, connection, and ownership.

The phrase "there is no such thing as a dumb question" is often used in public forums in an attempt to create a trusting environment such that *anyone can ask anything without fear of ridicule or criticism.* This makes sense, and, indeed, as leaders we must consistently strive to create psychological safety so that our people feel safe to ask questions when unsure or unclear.

That being said, some questions are better than others, and there are certain types of questions that leaders would do well to avoid asking.

After working as a psychologist for a quarter of a century, I am convinced that one of the most powerful influencing tools leaders have at their disposal is the effective use of questions, yet developing this skill set has not necessarily received the attention it deserves in leadership development programs.

While many leaders have been schooled in the vagaries of basic 'open' and 'closed' questions, there are far more powerful techniques that can be taught and mastered relatively quickly, yet can dramatically impact team culture and performance.

This chapter will provide some insights into how leaders can rapidly develop their use of effective questions, but first, a brief experiment …

Please DO NOT answer the following question!

What is the capital city of France?

So what happened? If you are like most people (and possess basic geographical knowledge) your brain automatically sent the answer to your conscious mind (despite the prior instruction asking you NOT to answer the question). You became aware of the word PARIS.

Questions are powerful because they steer our conscious attention, and our brains are hard-wired to seek answers to questions. It becomes vital then that leaders are *mindful* of the questions they ask. For example, consider the following two questions:

1. "Who is to blame?"
2. "What is just one thing we can do to move forward?"

Question 1 will likely result in the creation of fear and defensiveness, whereas question 2 is more likely to lead to a solution focus.

As leaders, being mindful of the types of questions we ask can quickly result in more helpful and productive responses from the teams we lead, as well as positively impacting the prevailing culture.

ASSUMPTIVE QUESTIONS

Again, consider the following questions:

1 "Are there any questions?"
2 "What questions do you have?"

Rookie facilitators and trainers commonly ask question 1 at the end of a training session. The question is closed and non-assumptive, and a clear response option is simply 'no!' (Usually evidenced by the group sitting silently in front of the facilitator and wondering if they can leave).

In stark contrast, question 2 is open and *assumes there are answers* and is therefore more likely to elicit responses from the group.

Other assumptive questions for the above scenario include:

- "What is the main question you have, based on what we've covered so far?"
- "At this point in the session, what do you think is the most frequently asked question?"

Questions like these can help engage our teams during the start of shift meetings and toolbox talks.

Assumptive safety questions such as "What do you see as the greatest risk for the oncoming crew?" are extremely powerful and are much more likely to elicit helpful responses than passive questions such as "Are there any risks?"

ASSUMPTIVE VERSUS LOADED QUESTIONS

There is a big difference between *assumptive* and *loaded* questions, in terms of both the intent and the results. Assumptive questions are not seeking a specific answer and are used based on the premise that people have a contribution to make, and to make it more likely they will voice their thoughts.

Loaded questions, on the other hand, are often used as a rhetorical tool that seeks to limit the answers to the questioner's self-serving motives.

Examples of loaded questions include:

- "Don't you think a smart person would know that?"
- "Haven't you realized that's wrong yet?"
- "Why is your team so lazy?"
- "Don't you agree that it could be done faster?"

They often begin with question stems such as:

- "Given that ...?"
- "Since we already know ...?"
- "Why do people always ...?"
- "What is the sense in ...?"

Loaded questions often contain unjustified assumptions and are manipulative in nature; hence they erode trust – in short, stop asking them!

INTERNALLY LOCUSED ('ABOVE THE LINE') QUESTIONS

For a number of important reasons it is desirable for leaders to encourage an internally locused (often more colloquially referred to as 'above the line') mindset within their teams. Internally locused teams tend to perceive more personal control over situations and are more likely to take positive action. In contrast, externally locused thinking tends to take our focus away from personal control and can result in inertia, blame, justification, and denial. Moreover, people with an internal locus of control are more likely to speak up, whereas individuals with an external locus of control are more likely to remain silent (Kahya, 2015). Hence, by fostering an internally locused ('above the line') mindset within their teams, leaders can help to increase empowerment, engagement, and psychological safety.

The types of questions leaders habitually ask their teams can exert a strong influence (positive or negative) on the prevailing locus of control within their teams. The basic rule here is that *internally locused questions will tend to elicit internally locused answers.* Equally, externally locused questions will encourage externally locused answers.

EXAMPLES OF EXTERNALLY LOCUSED ('BELOW THE LINE') QUESTIONS

- "Who is to blame?"
- "Why do I have to supervise everything you do?"
- "Why are we so unlucky?"
- "What's the point?"

Such questions can be very damaging to relationships, morale, and the culture in general. They serve no useful purpose yet often result in the team feeling helpless and focusing on blame and negativity.

EXAMPLES OF INTERNALLY LOCUSED ('ABOVE THE LINE') QUESTIONS

- "What are we learning from this?"
- "What will we do differently next time?"
- "How do you think we can best move forward with this?"
- "What are your thoughts on how we can improve this?"

Such questions – especially when used consistently by leaders – lead to a solution focus as well as a greater sense of control and empowerment within their teams. Moreover, just the invitation to be involved in decision-making and problem-solving helps teams to experience a sense of belonging and engagement, which in turn fosters self-efficacy, psychological safety, and trust.

A word of caution.

While it is generally more productive to focus on positives, there are exceptions. For example, don't ever do a dynamic risk assessment as an optimist! Sometimes it pays to be a pessimist, particularly when it comes to hazard identification (a pre-mortem is always better than a post-mortem!).

QUESTIONS THAT PROMOTE A SENSE OF UNEASE

High reliability organizations (HROs) – generally associated with high hazard industries such as nuclear and aviation – are renowned for possessing a sense of 'chronic unease' when it comes to thinking about potential risks. Despite inhabiting the upper stages (proactive and integrated) of safety climate maturity, leaders in HROs would not say, "We haven't had any incidents, we are doing great – let's have a barbeque!" Rather, they would reflect on questions such as "What are we missing?" "What else do we need to do?" and a constant refrain of "What if …?"

The reflexive tendency for culturally mature organizations to ask such questions promotes a state of 'collective mindfulness,' especially when combined with high levels of psychological safety that enables team members to speak up freely. These attributes are hallmarks of integrated safety cultures.

Unfortunately, many (if not most) organizations are yet to create the level of trust required for their people to feel at ease when sharing bad news. Rather than chasing unachievable, glib, and binary aspirations such as 'zero harm,' I believe a much more compelling, worthy, and attainable goal for all organizations would be to create the climate necessary for the workforce to feel unequivocally safe about speaking up when something is not right. The next two chapters offer powerful strategies to assist leaders in realizing this goal.

KEY POINTS

- The brain is hard-wired to answer questions.
- Leaders can use targeted questions to steer attention and invite their teams in.

- Above the line questions are more likely to elicit above the line, solution-focused answers.

REFLECTION QUESTIONS

- On balance, when communicating with your teams, do you do more telling or asking?
- What is an example of an occasion when you asked your team loaded or 'below the line' questions? What would have been a better question in that situation?
- What are some examples of effective 'above the line' questions you could ask your team during your next meeting?

8

DON'T SHOOT THE MESSENGER! MAKING IT SAFE TO SHARE BAD NEWS

Edmondson (2019) stresses the need for teams to feel psychologically safe enough to share bad news. Additionally, she points out that activities inviting teams to share candid feedback ought not to be a 'hit and miss' affair, but are most valuable when structurally embedded within the company's day-to-day operations.

In reality this rarely happens. More often than not a leader encounters a challenge and seeks solutions based on her/his own experience and/or input from other members of the leadership team. The solutions are then merely imposed upon the teams.

There are a number of challenges and wasted opportunities associated with this approach, and I see the above play out frequently in terms of the development of safety policies and procedures.

Consider the following costs:

- Without input from team members, the leader is assuming the work 'on the ground' is done as she or he imagines, which could well be different to how work is actually done.
- If the solutions are merely imposed upon team members without consultation, they will have less ownership of them and will be unlikely to fully engage with the changes.
- Without input from the end users, the solutions, quite simply, will not be as relevant and fit for purpose as they could be.
- Opportunities to 'bring in' the teams and build trust, ownership, and psychological safety are lost.

INVITING YOUR PEOPLE IN

After a company-wide rollout of the *Care Factor Program*, we assist our client organizations to form a group of 'champions.' This group of people (usually volunteers drawn from the workforce after demonstrating a high degree of enthusiasm during the program rollout) meets with leaders once a month to share feedback on where the new initiatives are gaining traction, as well as highlighting any identified areas of challenge. This regular, structured meeting helps embed positive gains while also providing a regular opportunity to increase the flow of authentic two-way communication. In *The Fearless Organization*, Edmondson (2019) describes how *Pixar* (among other corporations) utilizes a similar approach to elicit candor from staff by forming a group they call the 'brain's trust.' There are, of course, mutual responsibilities and expectations that are followed to ensure such candid feedback is expressed and received in respectful ways. Input from the group needs to be constructive rather than personal or disparaging. Similarly, leaders need to listen without defensiveness, understanding that the group's feedback is benevolent in nature and derived from a sincere intent to help.

Organizations need not limit such invitations to arbitrary groups, as described above. In my experience, leaders who regularly *bring their own teams in* to share frank and forthright feedback enjoy a number of worthy rewards:

- Teams learn it is safe to share ideas, challenges, and even bad news.
- Teams have the knowledge they actively contribute to resolving challenges.
- Increased team esteem, pride, empowerment, and satisfaction.
- Increased perception of personal control and influence.
- Increased trust and psychological safety.
- The team just works better.

A good deal of our time with leaders is spent coaching them to be more able to facilitate such regular and structured meetings. The process we most frequently utilize in these sessions is a relatively simple one, yet – when facilitated well – is an extremely powerful activity that engages our teams and invites them to identify areas requiring change. Moreover, the process provides a psychologically safe forum for teams to subsequently collaborate with leaders to identify solution-focused pathways, while simultaneously promoting a culture of trust, inclusion, and ownership.

This tool is the *Stockdale Paradox*.

THE STOCKDALE PARADOX

Background

Admiral James Stockdale was the highest-ranking naval officer to be held in the infamous 'Hanoi Hilton' prison camp during the Vietnam War.

Stockdale and his men were part of the so-called Alcatraz gang – American prisoners who were frequently held in solitary confinement and routinely subjected to various methods of psychological and physical torture, with Stockdale himself being tortured over 20 times during his captivity.

Despite this, Stockdale maintained a mindset of stoicism and resilience. He accepted the 'brutal facts' of his situation, and, rather than burying his head in the sand by moving into denial, pretense, or avoidance, he focused his energies on what he *could* control and influence and did whatever he could to lift the morale and prolong the lives of his men.

After Stockdale's release and subsequent return to the USA, Stockdale was interviewed by renowned psychologist and author Jim Collins and shared his perspectives on how he and his men had dealt with such a harrowing ordeal. During the interview, Collins asked Stockdale what was different about the men who *didn't* make it out of the camp, and Stockdale's response was somewhat surprising and counter-intuitive – he said, "That's easy – they were the optimists!"

He clarified as follows:

> They were the ones who said, "We're going to be out by Christmas." And Christmas would come, and Christmas would go. Then they'd say, "We're going to be out by Easter." And Easter would come, and Easter would go. And then Thanksgiving, and then it would be Christmas again. And they died of a broken heart.

He then added:

> This is a very important lesson. You must never confuse faith that you will prevail in the end – which you can never afford to lose – with the discipline to confront the most brutal facts of your current reality, whatever they might be.

The Stockdale Paradox

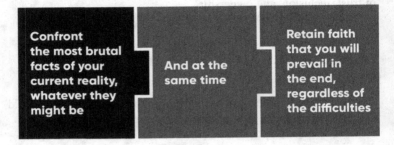

Figure 8.1 The Stockdale Paradox.

Collins went on to describe this powerful philosophy of duality in his classic work *Good to Great* as the *Stockdale Paradox* (Collins, 2001) (Figure 8.1).

Here is the paradox: While Stockdale unfailingly sustained the belief that he and his men would come through, he found that it was the most blindly optimistic of his fellow POWs who failed to survive. By denying the harsh reality of their situation, the optimists failed to confront the most *brutal facts*, instead of relying on avoidance, denial, or wishful thinking.

The *Stockdale Paradox* describes the *realistic* optimism that is essential to maintaining resilience, by looking at a challenge objectively and acknowledging the brutal facts while holding onto the confidence that comes from knowing we get to choose how we respond.

USING THE STOCKDALE PARADOX WITH YOUR TEAM

What Does This Process Look Like?

In my experience, it is desirable to facilitate a Stockdale Paradox activity regularly to resolve challenges as they arise. Moreover, when the activity becomes a routine component of a group's schedule, the team becomes increasingly skilled at deriving the maximum benefit from the process.

The three-step activity does not have to be overly long, although naturally more complex challenges require more time to work through.

Generally, we'll start with two large sheets of flip-chart paper on the wall. On the first sheet we'll give the heading *Our Brutal Facts*.

Step 1: Identifying Brutal Facts

On this sheet, participants record what the team identifies as current challenges. The facilitator helps the team to get specific about the challenges, as the clearer they become, the more powerful the second part of the activity can be.

If being invited to share bad news is relatively new to the team, they may be a little hesitant at first, particularly if trust is low. If this is the case, it is important the facilitator makes the process safe by stressing that the whole objective of this part of the exercise is to identify areas for change and that the team's involvement is welcome and valuable. Importantly, the facilitator needs to avoid being defensive, argumentative, or belittling as the responses emerge – such a reaction can immediately shut the team down and actually damage trust in the process.

When a challenge is identified, acknowledge the team and see if you can encourage them to provide more information. Often the initial 'brutal fact' is quite broad; for example, "Some contractors don't speak up about safety issues" (this is not an infrequent example). See if you can gain a little more information by using encouraging statements and gently probing questions such as, 'that's interesting, can you tell me more about that?'

Further information could include, for example,

Some are worried that if they speak up, they'll be fired – as they are casual workers that can easily happen. I think some of the contracting companies are also concerned that if too many near misses are reported, the client will get rid of them.

(again, a fairly common response)

Now the 'brutal facts' are really coming into focus. I'll sometimes go a step further with questions like, "So what? What could happen as a result of this? What makes this challenge so brutal?" Inquiries of this nature can elicit responses such as, "If they are hiding near misses they can't learn from them, and it's likely someone is going to get hurt."

Now we are getting really clear, and the facilitator needs to acknowledge the team for their candor and authenticity – it lets the team know the process is safe, that they are being heard, and that they can raise bad news without fear of retribution.

Step 2: Control and Influence

The next part of the process involves teaching (or reminding) the team what they can control and/or influence (as well as what they can't). This is usually done by moving through the circles of control, influence, and concern popularized by Covey (2004) (Figure 8.2).

The *circle of concern* may seem to align to some of the team's perceived brutal facts (i.e., things they believe they can't control nor influence); nevertheless, as the model suggests, *we can always control how we respond* (i.e., our *thoughts*, *feelings*, *words*, and *actions*).

Next we ask the team to return to their identified brutal facts and indicate for each aspect of the challenge whether they think there are elements they can control, influence, both, or neither. For example, teams will (rightly) conclude that they can't control whether a contractor speaks up or not, but they may well believe they can influence him/her.

Importantly, this process is never about letting the leadership group off the hook, as something else that team members frequently identify as being within their control is providing management with feedback on what leaders need to take responsibility for and/or change.

Step 3: Our Chosen Responses
Once the team is clear about what they can control and/or influence, we move to the second sheet of paper that has the heading 'Our chosen responses.'

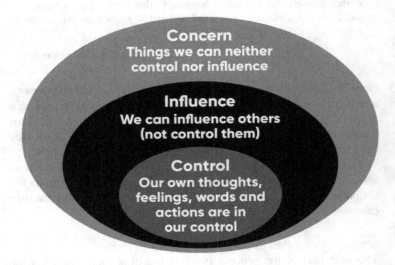

Figure 8.2 Circles of control, influence, and concern.

On this sheet the team writes down specific actions resulting from what they have agreed they can control and/or influence. The language needs to be specific and action-oriented. For example, if they have agreed they can influence a contractor to speak up, what would that ability to influence look and sound like in practice?

If they have agreed they can influence their management team to act upon the need to build trust with a contractor, how would they apply their influence? Hence, the content from sheet two is then operationalized as a clear plan of action (who, what, where, when, etc.).

The Stockdale Paradox activity is a fairly simple process; however, when facilitated well (and regularly) it has some enormous payoffs:

- It invites the team in.
- It teaches the team that sharing 'brutal facts' (bad news) is safe.
- As the process helps teams to get clear about what they can control and influence, it builds optimism and resilience.
- It creates a sense of ownership.
- It increases the flow of authentic communication ('top down' and 'bottom up').
- It builds trust.

Without bringing in their teams, leaders will struggle to fully understand the differences between 'work as done' (by teams on the ground) and 'work as management imagines it is performed.' Moreover, team involvement means it's less likely that workers will feel like organizational changes and revised safety policies are merely forced upon them – rather, they perceive ownership and inclusion, and the result is increasing engagement and psychological safety.

USING THE STOCKDALE PARADOX WITH YOUR LEADERSHIP TEAM

The Stockdale Paradox activity need not be restricted to use with work groups. Board members, senior leadership teams (SLTs), and frontline leaders can all gain great insights from working through their own (and others') *brutal facts*.

Your responses to the *reflection questions* at the end of each chapter will form a useful summary that you could use as the basis for a *Stockdale Paradox* activity with your leadership group. Any areas you identified as requiring change (within yourself, your leadership team, or the organization) could be written up as 'brutal facts,' permitting the leadership group to better understand what they can control and/or influence and to subsequently develop an action plan.

KEY POINTS

- It's crucial our people know it is safe to share bad news.
- A team's willingness to share bad news will depend upon the presence of psychological safety.
- Forums that invite our teams to contribute need to be structurally embedded within the company's day-to-day operations.
- The Stockdale Paradox provides the basis for a powerful activity to facilitate the above.

REFLECTION QUESTIONS

- Do you (or other leaders in your company) ever 'shoot the messenger?'
- How often do you invite your teams in? Do you currently have a 'champions' group or 'Brain's trust?'
- To what degree are employees asked to contribute to the design of safety tools, systems, and processes?
- What is one action you could take to move toward a regular, embedded forum that invites your team in?
- What will be the likely payoffs when you take that action?

While we're on the subject of sharing bad news, let's now turn our attention to some of the most brutal facts that leaders can ever be faced with – the aftermath of a workplace incident.

9

RESPONDING AFTER AN INCIDENT

Getting defensive doesn't hide the fact that you know you could have done better. Stop putting your energy into your excuses.

– Tony Curl

Prior to me sitting down to write this chapter, news came through of another mining fatality, this time in Western Australia. The latest tragedy follows a spate of mining deaths in Queensland's Bowen Basin over the last two years. Company press releases from such tragic events all contain variations of the following line: "The safety of our people is always our highest priority."

Whether or not that statement is accurate will likely be confirmed on completion of subsequent inquiries, but now is not the time for such claims. Just let us know how the primary and secondary victims are and that all impacted by the incident are receiving the care and support they need. Reciting platitudes such as "safety is our highest priority" at these times comes across as defensive and political and actually takes the focus away from where it needs to be – care for the victims. I often wonder if these companies all use the same lawyers when drafting post-incident press releases!

How a company responds after a significant incident can reveal a great deal about its culture. On occasion, even the more culturally mature organizations can regress by displaying panicked reactivity and subsequently incur a loss of trust from their staff and reputational damage in the eyes of their stakeholders.

THE DREAMWORLD TRAGEDY

'Dreamworld' is an iconic theme park on Australia's Gold Coast (which incidentally is my home town). On October 25, 2016, four people lost their lives after a malfunction on the Thunder River Rapids ride. The deaths rocked many Australians. The Gold Coast is a tourist mecca for many Australian and overseas visitors, and a visit to Dreamworld was often a much-anticipated event on their itinerary. Many of us remember having fun on that very ride, and, hence, there was an emotive realization that this tragedy could have happened to anybody's family, including my own.

I vividly remember watching the subsequent press conferences and that the message I was receiving from the company's executives seemed more about maintaining the organization's reputation rather than expressing genuine care for those affected by the incident. Sure, all the predictable banalities were there, about safety being the highest priority, and how the company was looking after the affected families; yet at parent company Ardent Leisure's annual general meeting (held just days after the tragedy) it was revealed that executives had *not* contacted the families of the four people who died, despite claiming they had!

In a tense press conference, Ardent Leisure's CEO, Deborah Thomas, claimed she had not called the family of two of the victims because she "didn't know how to contact them."

Challenging claims that there was an unsafe culture at the theme park, Chairman Neil Balnaves claimed that "Dreamworld has a strong safety culture and this is of paramount importance to the Board – it is not to be underestimated."

The coroner, however, came to some very different conclusions, labeling safety practices at the theme park "irresponsible, dangerous and inadequate," and stating that safety failures at Dreamworld were systemic.

In other astonishing findings, it was revealed that no risk assessment had been undertaken on the Thunder River Rapids ride in its 30 years of operation and that the ride had broken down five times in the seven days prior to the incident!

Such discoveries led the coroner to conclude that "such a culpable culture can only exist when leadership from the board down are careless in terms of safety."

Perhaps it's just a fear-driven reflex – human nature – for senior leaders to react with defensiveness after such a serious incident. When the two recent Boeing 737 MAX crashes killed hundreds of people, Boeing executives were quick to leap into defense mode and raise pilot error as a likely cause, which subsequent inquiries found not to be the case.

I recall an occasion where I had been requested to assist the General Manager (GM) in a debriefing process after a fatality on a mine site. The work area where the incident occurred had been made safe, and the family of the deceased worker ('Doug') had been notified. When I arrived at the site, I was taken into a room where the GM was sitting with five employees who were members of the same team as the deceased. None of the five team members had witnessed the incident, but all were close friends of Doug.

I immediately noticed that the GM was doing all the talking, and the team members were looking at the floor. Most of the five employees had their legs crossed and their arms folded across their chests (closed off from the GM's messages). I then tuned into the GM's words. He was saying things like: "I don't understand, we keep telling people to stop the job if things are not right. There's never a need to take shortcuts! We just reviewed the procedures … I don't get it!"

I'm not sure if he was trying to convince himself or his employees; however it was crystal clear that the team members weren't interested in what he was saying – they continued to stare at the floor. I intervened by simply asking, "Guys, what is it you need the most right at this moment?" All five immediately looked up and made eye contact with me, and then one employee said, "I just want to go and see my wife – she'll have heard about the incident and she's close to Doug's wife – she'll be beside herself." Several of the other employees followed his lead, nodding and sharing similar sentiments such as, "Same with me, I want to go and see Doug's family and see how I can help."

Picking up on my approach, and the rapidly changing dynamics in the room, the GM said,

> Absolutely understandable – just let us know if there's anything you need to say or do before you leave site. I'll call you later in the day to see how you're all doing, but if there's anything you need please call me directly on my mobile.

Again, the GM's initial, reflexive reaction was to go into defense mode – that is understandable. To be fair, he was also in shock and was no doubt feeling fearful about potentially negative impacts on his own and the company's reputation. While understandable, defense and blame at this point are unhelpful and potentially very damaging. Quite simply, the singular focus at such times needs to be on providing care and support for the victims.

Mercifully, most workplace incidents do not result in death or serious injury. Slips, trips, and falls remain the most common industrial accidents, often incurring lower levels of physical harm. Nevertheless, as the above examples illustrate, how employees (and secondary victims) are treated following even relatively 'minor' workplace incidents can have significant implications for both the individuals concerned and the organization's culture and reputation (Heraghty et al., 2020).

A RETRIBUTIVE JUST CULTURE

Wherever you find a problem, you will usually find the finger-pointing of blame

– Stephen R. Covey

[Note: Professor Sidney Dekker (2020) has made available a "Restorative Just Culture Checklist," and I highly recommend interested readers download the resource which Professor Dekker has kindly placed in the public domain. A copy can be found in the Appendix (along with the website address for downloading the original document). Some key points from the checklist are discussed below].

Dekker (2020) suggests that after an incident, leaders of organizations who demonstrate a *Retributive Just Culture* ask questions such as:

1. What rule is broken?
2. How bad is the breach?
3. What should the consequences be?

In my view, such questions lead to *parent–child* processes and interactions, which are often characterized by the use of top-down, hierarchical language such as 'non-compliant,' 'offender,' and 'violation.'

When leaders become defensive after workplace injuries, the use of such language infers blame, which is often leveled at the operator(s). Such a *retributive* reaction inevitably leads to a loss of trust and probable damage to the company's reporting culture through reduced psychological safety. Where staff members are fearful of being blamed for adverse events, they can feel inhibited about speaking up and are less likely to tell the truth in the future (Kaur et al., 2019). As Dekker noted,

a retributive just culture is linked with hiding incidents and an unwillingness to report and learn. The more powerful people that are in an organization, the more 'just' they find their retributive just culture. A retributive response doesn't identify systemic contributions to the incident, thus inviting repetition.

(2020, p. 2)

A RESTORATIVE JUST CULTURE

In contrast, authentic, values-based leaders are able to navigate more mature courses of action post-incident. Their primary attention is focused on three key *restorative just culture* questions:

1. Who is hurt?
2. What do they need?
3. Whose obligation is that? (Dekker, 2020)

Such questions steer leaders toward the ethical engagement of stakeholders, enable emotional healing of those affected by the incident, and, ultimately, facilitate organizational learning and improvement (Kaur et al., 2019). Moreover, organizations that adopt a restorative approach are less likely to regress to the lower levels of safety climate maturity and more likely to maintain high levels of trust and psychological safety.

Few processes better illustrate the shift from *compliance to care* as the move from a retributive to a restorative just culture. Again, the approach is neither complicated nor technical. It merely requires the application of those rare commodities: genuine care, authenticity, and values-based leadership.

KEY POINTS

- How an organization responds after an incident can have a dramatic impact on trust and psychological safety.
- Defensiveness and blame are common in apathetic and reactive cultures.
- Blaming a worker after an incident leads to reduced trust and a subsequent reduction in people's willingness to speak up.
- A *Restorative Just Culture* approach helps leaders to respond ethically and demonstrate care to those affected by an incident, protecting the psychological safety of the team.

REFLECTION QUESTIONS

- How does your organization typically respond after an incident?
- Would the typical responses tend to resemble a retributive or restorative just culture?
- How do you believe you would have responded if you had been the GM in the above scenario?
- What steps could you take to influence positive change in your organization regarding post-incident responses?

The above discussion around the *Restorative Just Culture* approach, and its developer, Professor Sidney Dekker, provides a smooth transition into the final chapter – looking at the Safety Differently movement, of which Dekker is a leading figure.

10

DOING SAFETY DIFFERENTLY
From Compliance to Care

The secret of change is to focus all of your energy, not on fighting the old, but on building the new.

– Socrates

I first encountered the term 'Safety Differently' back in 2013 during a conversation with a participant while on a break from one of my Safety Leadership sessions. He said, "I'm meeting up with a group of leaders tomorrow to talk about doing safety differently. Based on what you've been sharing here today, you ought to come along – it's right up your alley."

In fact Professor Sidney Dekker coined the term in 2012 in the header of an email he was sending out to organizations he thought might be interested in breaking away from traditional safety approaches.

In the eight years since those initial meetings took place, the whisper has turned into a roar, and it would be rare indeed these days to attend a safety conference anywhere in the world without encountering at least one or two speakers covering the topic of *Safety Differently*.

Given the approach's current prevalence, I'm sure many readers would be familiar with its central tenets; however, for the relatively uninitiated a synopsis is presented below.

WHAT IS SAFETY DIFFERENTLY?

Traditional safety (that which Erik Hollnagel labeled 'Safety I') tends to view people as a problem to be managed. Based on this top-down model of leadership, safety policies, procedures, and systems are often imposed upon employees with little or no consultation.

In contrast, the *Safety Differently* approach fundamentally sees people as the solution, recognizing their skills, knowledge, and experience. Hence, employees are *invited in*, and as Dekker says, *Safety Differently* replaces control with curiosity, prescription with participation, and instructions with involvement. As such, *Safety Differently* makes the bold leap toward decentralizing power and decision-making about safety by involving frontline workers. It asks the people who actually perform the work how things ought to be done.

As most of my clients and colleagues know, I'm a strong advocate for doing *Safety Differently*, and the more I work with clients toward this worthy goal, the more I believe that making concerted efforts to build *adult–adult* relationships and psychological safety helps organizations to evolve toward this approach organically. Conversely, if trust and psychological safety are low, it will be difficult to introduce a novel framework, as any new initiative (regardless of merit) is unlikely to gain traction among a fearful or untrusting workforce – the below quote from Chapter 1 rings in my ears and is worthy of repetition here.

> Unless the mistrust of the workforce can be overcome then even the most well-intentioned and sophisticated management initiatives will be treated with cynicism and undermined.
>
> Gunningham and Sinclair (2012)

Leaders seeking to embark upon a journey toward doing *Safety Differently* would do well to consider whether or not they have first created a climate in which the approach is likely to be embraced at the sharp end.

In my opinion, a successful *Safety Differently* initiative is *entirely reliant on trust*. First, leaders themselves must take a giant leap of faith in trusting that if they allow their workers to become fully involved in co-designing safety processes, the resulting changes (for example, in risk assessment activities) won't lead to disaster! Equally, if the workforce has a strong mistrust of management (perhaps based on years of working within a top-down, hierarchical parent–child culture), suddenly being invited to run the show could easily elicit suspicion, cynicism, and a middle finger!

Despite the role of trust being foundational to the approach, it has received scant attention in the *Safety Differently* literature and discussion – I believe this needs to change. If organizations fail to lay the requisite foundations for successful implementation, the worthy

approach could become just another safety fad. With interest in *Safety Differently* running at fever pitch (at least within academia), overzealous leaders could easily allow their initial fervor to drive a premature implementation. My suggestion would be ...
First create trust!

When leaders have the requisite trust in their teams and have established psychological safety throughout their organization, *Safety Differently* holds great promise. Moreover, where such trust already exists, the very nature of the *Safety Differently* approach is likely to help further embed and sustain psychological safety. I believe leaders who master the skills and implement the strategies I have outlined in the previous chapters will find that their journey toward *Safety Differently* has already begun. Once trust and psychological safety are established, doing *Safety Differently* then becomes a matter of leaders consistently tapping into the 'wisdom of the crowd.'

HOW DO WE START?

On the one hand, Dekker states that *Safety Differently* is not prescriptive and that there are no checklists to follow in order to successfully implement the approach. In fact he suggests that dogmatically following a recipe would indicate a regression to traditional safety thinking. Hence, many leaders who intuitively find the philosophy of *Safety Differently* appealing may be disappointed at the lack of a *Safety Differently User's Manual* and beat a hasty retreat to the comfort of *Safety I*.

On the other hand, as highlighted in the previous chapter, Dekker offers a highly prescriptive approach to facilitating a *Restorative Just Culture* – a central element of the *Safety Differently* approach. This apparent contradiction arises because *Safety Differently* is centered around leaders inviting their people in, and therefore the output from such a collaborative methodology will necessarily be unique to a particular leader and his/her team – it is not a *one-size-fits-all* approach.

Nevertheless, contributors to the *Safety Differently* literature have identified several core processes that – while not constituting a recipe – may act as a foundation for successful implementation.

These core processes include:

- Decentralizing and devolving power
- De-cluttering

- An analysis of *work as imagined* versus *work as done*
- Appreciative investigations
- Restorative Just Culture

Moreover, as early adopters share their first tentative steps toward operationalizing *Safety Differently*, some common themes have started to emerge to guide aspiring agents of change. While successful implementation will still require leaders to think things through and to bring a curious mindset to interactions with their teams, that need not prohibit learning from the pioneers who have successfully put *Safety Differently* processes into action.

NEXT GENERATION SAFETY LEADER PROFILE 2: SIMON BOWN

In researching this chapter I spoke with Simon Bown, who was Head of Health, Safety, and Environment at London Luton Airport. He recently shared an article outlining the steps he had taken to introduce a *Safety Differently* approach at the airport. We'll have a glimpse into his journey shortly, but first let me share my thoughts about Simon as a leader (and indeed, a person).

It became obvious to me early in our conversation that he is an authentic leader – he drops his masks! As well as being extremely knowledgeable about risk, hazard mitigation, safety systems, and other technical aspects of safety, Simon is someone I immediately felt at ease with. It was clear he wished to change things for the benefit of his team and his organization, rather than for any personal kudos. He had implicit trust in his team members and their knowledge, skills, experience, and ability to contribute. I would also bet a year's salary he had the trust *of* his team!

I'm sharing my perceptions of Simon to demonstrate a key point. If he was *not* an authentic, caring leader who had earned the trust of his team, as well as having trust *in* his team, it is likely that his well-intended approach would have fallen flat. However, given he *does* possess the attributes I described above, it was always going to be a safe bet that implementing *Safety Differently* methodologies would not only be well received but would further embed trust levels and maintain a high degree of psychological safety within his team.

Safety Differently is, by its very nature, an ideal approach for building trust, engagement, ownership, and psychological safety, and these are the reasons I'm an advocate. Yet without the key characteristics I ascribe to Simon, leaders will struggle to realize the full potential of the approach. As a case in point, without some prior intensive coaching, it would be highly unlikely that 'Nigel,' the surly HSE Manager from Chapter 1, or 'Bill,' the 'old school' Superintendent featured in Chapter 4, would have the requisite level of trust within their teams for a successful *Safety Differently* implementation.

How do we start? First create trust!

Simon is very generous with his time, and not only has he permitted me to share much of his article, but he recently updated and edited the text for inclusion in this book. So here is London Luton Airport's journey from traditional safety to doing *Safety Differently* in Simon's own words.

Case Study: Doing Safety Differently at London Luton Airport

One day I heard REM's 'Losing My Religion' on the radio and thought 'that's me with safety!' It was just after this point that I met with some truly inspirational individuals, first Daniel Hummerdal, then John Green, and finally Sidney Dekker.

It was after a masterclass and plenty of discussions with Daniel that I knew I was going to try my interpretation of Safety Differently, but that all-too-popular question was present – "Where do I start?"

Here goes!

Selling the new approach was first up. I met with my boss Nick Barton (the CEO), who listened intently and questioned rigorously. I think the main 'penny drop' moment for most is when asked the following question: Is it progressive for a business to only learn after a negative outcome event?

I would regularly challenge colleagues in meetings about not only focusing on negatives, but also learning from successful outcomes, empowerment, and innovation. The seeds were starting to sow! I wanted to plant these seeds before going live to the business.

If we were to successfully decentralize, devolve, and de-clutter, there was some additional work to be done. If I was to ask employees to self manage risk then I had to ensure that everybody had a base level of dynamic risk management skills. This is where I engaged with Justin Hughes from Mission Excellence, who used skills forged as a fighter pilot and a member of the Red Arrows to roll out some dynamic risk management under pressure training.

I had to build trust in my team so we were viewed as 'enablers' and not 'constrainers.' This led me to change the behaviors of my team and renaming them 'HSE Coaches' with the sole intention of coaching, mentoring, developing, and facilitating. In many places it is habitual to blame the person involved in any accident or incident without looking deeper for the drivers which influenced that person's course of action. This is where you usually find the real root cause, which is often an organizational failure. Traditionally safety people look to 'detect and correct,' but I tried to re-program the coaches to view variation as an insight instead. Follow three simple words when viewing variation: Ask, listen, and understand. The result was a success, with the coaches viewed more positively and trust being built.

I started the roll out in our Cargo Centre by presenting on the broad principles of Safety Differently, which progressed to appreciative investigations and then going out and doing some with the team. In our first batch, we uncovered some great finds with operatives being put at risk and not working to the agreed safe systems, as they said it was not possible to do so. This particular scenario was the transportation of coffins in a confined trailer. The operative would need to climb in beside and use fingertips to move the coffin so that the other operatives could maneuver the coffin out at the other end. The outcome of this is a redesigned trailer (by the operatives), which is totally fit for purpose. The Cargo staff said that the only way historically that this would have been resolved would have been if there was an accident or incident, and this would have probably resulted in somebody being disciplined. From this point the Cargo staff were very enthusiastic when seeing the benefits of proactively looking at their tasks, most importantly with their input as the expert.

Another interesting find was when a 'work as imagined versus work as done' review was undertaken. This task involved the decanting of large cargo pallets to split the loads and reallocate to other loads. This entails the use of a lowering work platform. We were about ten minutes in when one of the operatives asked if we could stop. I asked why, and he said, 'we don't even do it like this.' I asked why and he said they do it off the floor, as there is a small falling risk where the platform lowers and creates a gap between the pallet and pit, which cannot be guarded without affecting the operation of the platform. We talked through the actual 'work as done' and agreed that this approach had different risks but no higher than the work as an imagined way. A new safe system of work was developed and implemented only for a different team to say they preferred the original method. I then made the decision to empower the team to undertake the task in either of the agreed methods as they are the experts and are competent enough to choose how they undertake that task. This was our first dual system task at the airport, which is working very well.

When approaching the first revision of the H&S policy, I only subtlety introduced a few of the theories as I did not want to bombard the workforce with another 'management fad,' which they felt may disappear in time. I did this so that, when I produced the second more defined policy, people would look and think 'we already do a lot of that.' This proved to be a great success. I've just prepared version three, which I would say is practically the finished article. When defining these policies I needed to come up with a strap line which people could associate with, so we opted for: "Safety is not about the absence of accidents, it's the presence of trust, ownership, engagement and positives."

Another area that needed to be changed was our KPIs. Historically they were all negative outcomes such as RIDDORs, LTIs, and number of accidents, so in year one we agreed to split these 50/50 and year two went 100% positive outcomes. I generally devise this from the output of the employee survey as I ask specific questions to ascertain what is important to the employees and then devise appropriate KPIs from the feedback.

One of the biggest hurdles to overcome was transitioning from the practice of habitually blaming the person when anything happened, as do plenty of businesses that I've come across. If I had a pound for every time I've heard 'if only they had followed the safe system of work' I'd be a rich man! How about thinking along the lines of 'who actually comes to work with the intention of getting injured?' How about another deeper cause that it could be an organizational failure? Have we failed the person organizationally which drives and influences the individual to make a decision to get the job done?

This could be operational pressure, bonus related, target related, lack of resources, inaccurate process/procedures, incorrect equipment, etc. I needed to enable people to learn differently by not jumping to conclusions, talking to the frontline employees without a 'detect and correct' or blame mentality. I needed them to view variation as an insight and then let it be explained fully by following the 'ask, listen understand' methodology. This way the real expert was giving their perceptions and rationale of how they manage risk and the reasons why. We could also learn from positives by undertaking work as imagined versus work as done reviews, appreciative investigations, and learning reviews, which significantly contributed to a shift to becoming a learning culture from a blame culture. This also assisted in building trust at all levels as the fear of retribution was gradually diminishing by us following Sidney Dekker's restorative just culture model for all health and safety events.

Employees at all levels have enjoyed the empowerment, trust, and autonomy afforded to them and the ability to use their natural risk management skills while performing tasks that they are the experts in. There is no fear of retribution when one of the coaches is present as it's always a constructive conversation rather than telling them that they are doing something wrong or telling them how to do their job. Bureaucracy has been rationalized in some areas, so only what is needed is present.

Plenty of impact has been made just by the pure use of language as certain words carry stigma. Here are some of the changes: advisors to coaches, accident to the learning event, accident investigation to learning review, and audits to continuous improvement opportunities.

One addition that has had a great impact is that when undertaking a traditional accident investigation (learning review here), a work as imagined versus work as done is undertaken with various other operatives to understand if what the injured person was doing was normal practice across the board or an isolated incident. This has to date provided some great learning opportunities that would probably not been identified if this exercise was not added to the learning review.

One question that I'm always asked about this approach is 'what does the regulator think?' I can honestly say that every regulator that I've discussed this with has been very supportive, as contrary to popular belief; we don't just say, "crack on boys and make it up as you go along" and "we are not bothered about smaller accidents and injuries." We are still fully compliant with regulatory standards and requirements; we just go about meeting that compliance from a different viewpoint. We are not obsessive about continuous audits and inspections to find fault. We examine normal work to identify collaborative improvement, which meets those requirements. Regulators have actually disclosed that they are supportive of this approach as it takes collaboration and engagement to another level by using the skills and knowledge of the experts who actually perform the role. I was also challenged by many who said it was impossible to get the new ISO45001 standard using the Safety Differently model. I am very pleased to say that in October of 2018, we were recommended for accreditation with zero non-conformity actions raised against us. To our knowledge, this makes us the first UK Airport Operator to achieve the standard and one of the first businesses globally to get the standard using this model.

Below is part of the feedback in our report:

A positive approach to health and safety is clearly embedded within the company. This was demonstrated across all levels and functions. Collaboration of information discussed with the leadership team confirmed this. There is a unique philosophy of Safety Differently that is led from the top and cascaded down throughout the organization. The approach gives people autonomy, trusting their staff, listening to them, and engaging with them. The company is achieving its strategic goal of running LLA in a proactive, positive, safe, and responsible manner.

Positive findings from this visit include:

- Establishing, monitoring, and communicating objectives based upon the employee engagement survey.
- Communication channels and collaboration with interested third parties.
- Leadership commitment to achieving a positive health and safety culture throughout the organization.
- The company has fully reviewed its management system to meet the requirements of ISO45001:2018. This has been subject to full assessment during this visit and is confirmed as meeting the requirements of the standard.

As my time at London Luton Airport draws to a close, I'm immensely proud of the journey the airport is now on. I thank my team, all employees, and former CEO Nick Barton for having the trust and belief in me to take H&S in another direction.

Adapted from Bown, S. (2019). The London Luton Airport safety differently journey. Safety differently. https://safetydifferently.com/the-london-luton-airport-safety-differently-journey/

To me, Simon epitomizes the next generation safety leader. He assisted his team and his organization to move from *compliance to care,* not by *imposing* new policies, systems, or rules, but by inviting his team in, and bringing to life their new tagline, "Safety is not about the absence of accidents, it's the presence of trust, ownership, engagement and positives." He didn't require a *Safety Differently Users Manual*; instead he channeled his passionate desire for change into curiosity, creativity, and collaboration with his teams.

Simon's approach also highlights and partially resolves some of the more common "yes, but," and "what if?" concerns would-be early adopters of *Safety Differently* frequently voice:

"Yes, but my boss would never go for that."
"What if the regulator doesn't approve?"
"What if my team does poor dynamic risk assessments?"
"Yes, but what about achieving the standards?"

The voyage undertaken by Simon's team is not shared here as a prescription, but rather as one example of how a fundamental shift in thinking about our teams can result in our employees' motivation for safety involvement moving from extrinsic (imposed) to intrinsic (owned) – a recognition that people are the solution, rather than a problem to be managed.

At the heart of Simon's approach was trust – the trust he was required to *have* in his team, and the trust he created *within* his team in order for them to experience the psychological safety required to fully engage, accept responsibility, and contribute so well.

KEY POINTS

- Safety Differently represents a shift away from traditional, bureaucratic, hierarchical approaches.
- Safety Differently sees people as part of the solution rather than a problem to be managed.
- Power and decision-making are decentralized, with the workforce fully invited in.
- The Safety Differently approach is one that is likely to assist the further development of trust and psychological safety.

REFLECTION QUESTIONS

- What (if any) elements of the Safety Differently approach have you or your company already adopted?
- What do you see as some of the benefits associated with moving further toward a Safety Differently approach?
- What examples of 'yes, but ...' resistance to moving toward the approach do you have or anticipate hearing from your leadership team?
- Are the 'what ifs' based on legitimate concerns or resistance to change, more generally? Is it preferable to stick with traditional safety approaches? Why? (Or why not?)
- What could be your next steps toward a more inclusive, humanistic approach to safety leadership?

EPILOGUE

Much has been written about the differences between management and leadership. My own belief, in essence, is that we manage things, we lead people. Traditional safety approaches have sought to manage people via the imposition of rules, policies, procedures, rewards, and punishment, which inevitably requires a bureaucratic, hierarchical top-down management style hell-bent on forcing compliance. It is this very approach which has stymied development of the one thing organizations require the most in order to manage risk and foster a caring culture – trust!

Not everything will (or need to) change. The next generation of safety leaders will still need to focus on risks and their mitigation. They will still need to meet regulators' standards and be involved in designing safe systems and practices. The difference is they won't be doing these things to their people, they'll be doing them with their teams.

The next generation of safety leaders will recognize every meeting, every toolbox talk, and every conversation as an opportunity to build trust and demonstrate care – they will also possess the requisite knowledge and skills to achieve this.

They will do less 'telling' and more asking, knowing that their teams have most of the answers they seek. These acts of humble inquiry will help leaders to invite their teams in, and the result will be ownership, pride, intrinsic motivation, and trust. Such psychologically safe teams will organically begin the shift toward doing Safety Differently and having the confidence to share their ideas, concerns, and even their mistakes.

This book has reflected a great deal on what great safety leaders actually do, yet to me such actions are more a reflection of who they are: values-based, authentic people, with the courage and humility to be vulnerable, drop their masks and genuinely engage with their people.

I am not alone in my assertions that safety leadership requires generational change toward a more humanistic approach. A recent survey by the Acre Frameworks Advisory panel (made up of 96 senior health and safety professionals from over 20 different sectors) found that 89% of panel members rated non-technical skills as the most important, and 100% reported that technical skills were the least important when recruiting and developing health and safety leaders (Acre, 2020). The advisory panel went on to say that "we believe that there is a call to action to effectively define a more forward-thinking set of non-technical skills. With this common language we can reinvent the way we teach and upskill the next generation."

The advisory panel identified the following attributes as key for the next generation of Safety Leaders:

- To be constant questioners rather than knowers
- To become masters of emotional and relational intelligence rather than of technical expertise
- To inspire others to find a sense of purpose in the work they do and create intrinsic motivation rather than offer purely extrinsic rewards
- To reveal unseen opportunities and challenges by embracing uncertainty and vulnerability rather than attempting to engineer humanity out of work processes
- To be a servant leader rather than a figure of authority in a hierarchical relationship
- To listen loudly and coach rather than command and control (Acre, 2020)

With so many senior health and safety professionals responding to the call for action, clearly the journey from compliance to care is already gaining momentum.

Are you with us, dear reader?

APPENDIX

Professor Sidney Dekker kindly made the (below) Restorative Just Culture Checklist available in the public domain. It can be accessed at the following site:

http://sidneydekker.com/wp-content/uploads/2018/12/RestorativeJustCultureChecklist.pdf

RESTORATIVE JUST CULTURE CHECKLIST

Restorative Just Culture aims to repair trust and relationships damaged after an incident. It allows all parties to discuss how they have been affected, and collaboratively decide what should be done to repair the harm.

WHO IS HURT?

ACKNOWLEDGED:
NO   YES

Have you acknowledged how the following parties have been hurt:
First victim(s) – patients, passengers, colleagues, consumers, clients
Second victim(s) – the practitioner(s) involved in the incident
Organization(s) – may have suffered reputational or other harm
Community – who witnessed or were affected by the incident
Others – please specify:......................................

WHAT DO THEY NEED?

EXPLORED:
NO   YES

Have you collaboratively explored the needs arising from harms done:
First victim(s) – information, access, restitution, reassurance of prevention
Second victim(s) – psychological first aid, compassion, reinstatement
Organization(s) – information, leverage for change, reputational repair
Community – information about incident and aftermath, reassurance
Others – please specify:......................................

WHOSE OBLIGATION IS IT TO MEET THE NEED?

IDENTIFIED:
NO   YES

Have you explored the needs arising from the harms above:
First victim(s) – tell their story and willing to participate in restorative process
Second victim(s) – willing to tell truth, express remorse, contribute to learning
Organization(s) – willing to participate, offered help, explored systemic fixes
Community – willing to participate in restorative process and forgiveness
Others – please specify:......................................

READY TO FORGIVE?

NO   YES

Forgiveness is not a simple act, but a process between people:
Confession – telling the truth of what happened and disclosing own role in it
Remorse – expressing regret for harms caused and how to put things right
Forgiveness – moving beyond event, reinvesting in trust and future together

ACHIEVED GOALS OF RESTORATIVE JUSTICE?

ACHIEVED:
NO   YES

Your response is restorative if you have:
Moral engagement – engaged parties in considering the right thing to do now
Emotional healing – helped cope with guilt, humiliation; offered empathy
Reintegrating practitioner – done what is needed to get person back in job
Organizational learning – explored and addressed systemic causes of harm

Public Domain. By Professor Sidney Dekker—Griffith University, Delft University and Art of Work. sidneydekker.com

Figure A.1 Dekker's Restorative Just Culture Checklist (Page 1)

BACKGROUND OF RESTORATIVE JUSTICE

Restorative Just Culture asks:
- **Who is hurt?**
- **What do they need?**
- **Whose obligation is that?**

Accountability is *forward-looking.*
Together, you explore what needs
to be done and who should do it

An **account** is something
you tell and learn from

Retributive Just Culture asks:
- What rule is broken?
- How bad is the breach?
- What should consequences be?

Accountability is *backward-looking*,
finding the person to blame and
imposing proportional sanctions

An **account** is something
you settle or pay

WHY AVOID RETRIBUTIVE JUST CULTURE?

A retributive just culture can turn into a blunt HR or managerial instrument to get rid of people.
It plays out between 'offender' and employer—excluding voices of first victims, colleagues, community.
A retributive just culture is linked with hiding incidents and an unwillingness to report and learn.
The more powerful people are in an organization, the more 'just' they find their retributive just culture.
A retributive response doesn't identify systemic contributions to the incident, thus inviting repetition.

GUIDANCE FOR USE OF *RESTORATIVE* JUST CULTURE CHECKLIST

On the checklist, mark where you think you are, like so: or so:
Together, the marks reveal what you still need to do.

HURTS, NEEDS AND OBLIGATIONS

An incident causes (potential) hurts or harms. This creates needs in the parties harmed.
These needs produce obligations for the (other) parties involved.
Restorative justice allows parties to discuss their hurts, their needs and the resulting obligations *together*.
Incidents don't just harm their (first) victim(s). They also (potentially) harm the second victim, supervisors,
the organization, colleagues, bystanders, families, regulatory relationships and the surrounding
community. All these parties have different needs arising from the harms caused to them.
The checklist allows you to trace the harmed parties, their needs, and the obligations on them/others.

FORGIVENESS

Forgiveness is not a simple act of one person to another. Forgiveness is a relational process that involves
truth-telling, repentance and the repair of trust. It takes time. Trust is easy to break and hard to fix. Some
first victims may be unwilling or unable to forgive. Second victims can also have difficulty forgiving
themselves. Parties need to have patience and compassion, and may end up going separate ways.

GOALS OF RESTORATIVE JUSTICE

- *Moral engagement* can mean accepting appropriate responsibility for what happened, recognizing
the seriousness of harms caused, and humanizing the people involved. Incidents can overwhelm an
organization (e.g. a legal, reputational, financial, managerial issue). It is easy to forget that it is also a
moral issue: What is the right thing to do?
- *Emotional healing* aims to deal with feelings such as grief, resentment, humiliation, guilt and shame. It
is a basis for repairing trust and relationships.
- *Reintegrating* the practitioner expresses the trust and confidence that the incident is about more than
just the individual. Expensive lessons can disappear from the organization if the practitioner is not
helped back into the job, and letting them go tends to obstruct the three other goals. If you fire
someone, what have you fixed?
- Restorative justice is better geared toward *addressing the causes* of harm because it goes beyond the
individual practitioner and invites a range of stories and voices. Forward-looking accountability is
about avoiding blame, and instead fixing things.

Figure A.2 Dekker's Restorative Just Culture Checklist (Page 2)

BIBLIOGRAPHY

Acre. (2020). Co-creating the future of H&S leadership development. Acre Frameworks Advisory Panel. https://www.acre.com/thought-leadership/co-creating-the-future-of-h-and-s-leadership-devel opment (accessed February 9, 2020).

Avolio, B., Gardner, W.L., Walumbwa, F.O., Luthans, F., & May, D.R. (2004). Unlocking the mask: A look at the process by which authentic leaders impact follower attitudes and behaviours. *The Leadership Quarterly*, 146, 2–34.

Bandura, A. (1997). *Self-Efficacy: The Exercise of Control*. New York: Freeman.

Bown, S. (2019). The London Luton Airport safety differently journey. Safety Differently. https://safetydifferently.com/the-london-luton-airport-safety-differently-journey/ (accessed February 11, 2020).

Burns, C., Mearns, K., & McGeorge, P. (2006). Explicit and implicit trust within safety culture. *Risk Analysis*, 26(5), 1139–1150.

Carrillo, R. (2020). *The Relationship Factor in Safety Leadership*. London: Routledge.

Cavazotte, F., Duarte, C., & Gobbo, A. (2013). Authentic leader, safe work: The influence of leadership on safety performance. *Brazilian Business Review*, 10(2), 95–115.

Collins, J. (2001). *Good to Great: Why Some Companies Make the Leap and Others Don't*. New York: Harper Collins.

Collins, J., & Porras, J. (2002). *Built to Last: Successful Habits of Visionary Companies*. New York: Harper Business Essentials.

Conchie, S.M. (2013). Transformational leadership, intrinsic motivation, and trust: A moderated-mediated model of workplace safety. *Journal of Occupational Health Psychology*, 18(2), 198–210.

Conchie, S.M., Taylor, P.J., & Charlton, A. (2011). Trust and distrust in safety leadership: Mirror reflections? *Safety Science*, 49, 1208–1214.

Covey, S.M.R., & Merrill, R.R. (2006). *The Speed of Trust: The One Thing that Changes Everything*. New York: Simon and Schuster.

Covey, S.R. (2004). *The 7 Habits of Highly Effective People: Restoring the Character Ethic*. New York: Free Press.

Dekker, S. (2017). Zero vision: Enlightenment and new religion. *Policy and Practice: Health and Safety*, 15(2), 101–107. doi: 10.1080/14773996.2017.1314070

Dekker, S. (2020). Restorative just culture checklist. Download available at: http://sidneydekker.com/wp-content/uploads/2018/12/Restorati veJustCultureChecklist.pdf (accessed February 21, 2020).

Dirik, H.F., & Seren, I. (2017). The influence of authentic leadership on safety climate in nursing. *Journal of Nursing Management*, 25(5), 392–401.

Edmondson, A. (1999). Psychological safety and learning behaviour in work teams. *Administrative Science Quarterly*, 44(2), 350–383.

Edmondson, A. (2003). *Psychological Safety, Trust and Learning in Organisations: A Group-Level Lens* (Unpublished doctoral dissertation or master's thesis). Harvard Business School, Boston, MA.

Edmondson, A. (2019). *The Fearless Organization: Creating Psychological Safety in the Workplace for Learning, Innovation, and Growth.* Hoboken, NJ: John Wiley & Sons.

Eid, J., Mearns, K., Larsson, G., Laberg, J., & Johnsen, B. (2012). Leadership, psychological capital and safety research: Conceptual issues and future research questions. *Safety Science*, 50, 55–61.

George, W. (2003). *Authentic Leadership; Rediscovering the Secrets to Creating Lasting Value.* San Francisco, CA: Jossey-Bass.

Gunningham, N., & Sinclair, D. (2012). *WP 85 – Building Trust: OHS Management in the Mining Industry.* Canberra: National Research Centre for OHS Regulation.

Heraghty, D., Rae, A.J., & Dekker, S. (2020). Managing accidents using retributive justice mechanisms: When the just culture policy gets done to you. *Safety Science*, 1, 26. doi: 10.1016/j.ssci.2020.104677.

Hollnagel, E., Wears, R.L., & Braithwaite, J. (2015). *From Safety-I to Safety-II: A White Paper.* The Resilient Health Care Net: Published simultaneously by the University of Southern Denmark; University of Florida, USA; and Macquarie University, Australia.

Hopkins, A. (2005). *Safety, Culture and Risk.* Australia: CCH Australia Ltd.

Hopkins, A. (2009). *Learning from High Reliability Organisations.* Sydney: CCH Australia Limited.

Hopkins, A. (2012). *Disastrous Decisions: The Human and Organisational Causes of the Gulf of Mexico Blowout.* Sydney: CCH Australia Limited.

Hudson, P. (1999). *Safety Culture – Theory and Practice.* The Netherlands: Centre for Safety Science, Universiteit Leiden.

Hudson, P. (2001). Safety management and safety culture: The long, hard and winding road. In: Pearse, W., Gallagher, C., & Bluff, L. (eds.), *Occupational Health and Safety Management Systems.* Melbourne, VIC: Crown Content, 3–32.

Jeffcott, S., Pidgeon, N., Weman, A., & Walls, J. (2006). Risk, trust and safety culture in UK train operating companies. *Risk Analysis*, 26(5), 1105–1121.

Kahya, C. (2015). The relationship between locus of control and organisational silence: A study on the Turkish academicians. *IIB International Refereed Academic Social Sciences Journal*, 6(19), 1–18.

Kaur, M., De Boer, R.J., Oates, A., Rafferty, J., & Dekker, S. (2019). Restorative just culture: A study of the practical and economic effects of implementing restorative justice in an NHS trust. *MATEC Web of Conferences*, 273, 01007. doi: 10.1051/matecconf/201927301007.

Long, R. (2012). *For the Love of Zero: Human Fallibility and Risk*. Kambah, ACT: Scotoma Press.

Luthans, F., & Avolio, B.J. (2003). Authentic leadership development. In: K.S. Cameron, J.E. Dutton, & R.E. Quinn (eds.), *Positive Organizational Scholarship*. San Francisco, CA: Berrett-Koehler, 241–261.

Mayer, R.C., Davis, J.H., & Schoorman, F.D. (1995). An integrative model of organisational trust. *Academy of Management Review*, 20, 709–734.

Mearns, K.J. (2008). Organisational support and safety outcomes: An un-investigated relationship? *Safety Science*, 46(3), 388–397.

Nielsen, M.B., Eid, J., Mearns, K., & Larsson, G. (2013). Authentic leadership and its relationship with risk perception and safety climate. *Leadership and Organization Development Journal*, 34(4), 308–325.

O'Dea, A., & Flin, R. (2001). Site managers and safety leadership in the offshore oil and gas industry. *Safety Science*, 37, 39–57.

O'Dell, C., & Grayson, C.J. (1998). If only we knew what we know: Identification and transfer of internal best practices. *California Management Review*, 40(3), 154–174.

Perrow, C. (1984). *Normal Accidents: Living with High-Risk Technologies*. New York: Basic Books.

Rae, D., & Provan, D. (2020). Is adopting a zero harm policy good for safety? Podcast at https://safetyofwork.com/episodes/ep12-is-adopting-a-zero-harm-policy-good-for-safety (accessed April 4, 2020).

Reason, J. (1997) *Managing the Risks of Organizational Accidents*. Farnham, Surrey: Ashgate Publishers.

Read, B.R., Zartl-Klik, A., Veit, C., Samhaber, R., & Zepic, H. (2010). Safety leadership that engages the workforce to create sustainable HSE. *Paper presented at the SPE International Conference on Health, Safety and Environment in Oil and Gas Exploration and Production*, Rio de Janeiro, Brazil, 12–14 April 2010.

Rotter, J. (1973). Internal locus of control scale. In: P. Robinson & R.F. Shaver (eds.), *Measures of Social Psychology Attitudes*. Institute for Social Psychology. London: Academic Press Inc, Harcourt Brace Jovanovich, 53.

Seligman, M. (2002). *Authentic Happiness: Using the New Positive Psychology to Realize Your Potential for Lasting Fulfillment*. New York: Simon & Schuster.

Sherratt, F., & Dainty, A. (2017). UK construction safety: a zero paradox? *Policy and Practice in Health and Safety*, 15(2), 108–116. doi: 10.1080/14773996.2017.1305040.

Skowron-Grabowska, B., & Sobociński, M. (2018). Behaviour based safety (BBS) - Advantages and criticism. *Production Engineering Archives*, 20, 12–15.

Smith, S. (2007). Behaviour based safety? Myth or magic. *EHS Today*, September 2007. http://ehstoday.com/safety/ehs imp 75429/ (accessed February 9, 2020).

Smith, T. (1999). What is wrong with behavior based safety? *Professional Safety*, September 1999. https://miningquiz.com/pdf/Behavior _Based_Safety/whats_wrong_with_behavior_based_safety.pdf (accessed February 9, 2020).

Triplett, S., & Loh, J. (2017). The moderating role of trust in the relationship between work locus of control and psychological safety in organisational work teams. *Australian Journal of Psychology*, 70, 76–84. doi: 10.1111/ajpy.12168.

Weick, K., & Sutcliffe, K. (2001). *Managing the Unexpected: Assuring High Performance in an Age of Complexity*. San Francisco, CA: Jossey-Bass.

Weick, K., & Sutcliffe, K. (2007). *Managing the Unexpected: Resilient Performance in an Age of Uncertainty*. San Francisco, CA: Jossey-Bass.

Weick, K., Sutliffe, K., & Obstfeld, D. (1999). Organizing for high reliability: Processes of collective mindfulness. In: R.S. Sutton & B.M. Staw (eds.), *Research in Organizational Behavior*, Volume 1. Stanford, CA: Jai Press, 81–123.

Wuebker, L.J. (1986). Safety locus of control as a predictor of industrial accidents and injuries. *Journal of Business and Psychology*, 1, 19–30.

Zacharatos, A., Barling, J., & Iverson, R.D. (2005). High performance work systems and occupational safety. *Journal of Applied Psychology*, 90, 77–93.

Zwetsloot, G.I.J.M., Kines, P., Wybo, J.L., Ruotsala, R., Drupsteen, L., & Bezemer, R.A. (2017). Zero accident vision based strategies in organizations: Innovative perspectives. *Safety Science*, 91, 260–268. doi: 10.1016/j.ssci.2016.08.016.

Printed in the United States
by Baker & Taylor Publisher Services